The Paper Trail

ALEXANDER MONRO

The Paper Trail

An Unexpected History of the World's Greatest Invention

ALLEN LANE

an imprint of

PENGUIN BOOKS

ALLEN LANE

Published by the Penguin Group
Penguin Books Ltd, 80 Strand, London WC2R ORL, England
Penguin Group (USA) Inc., 375 Hudson Street, New York, New York 10014, USA
Penguin Group (Canada), 90 Eglinton Avenue East, Suite 700, Toronto, Ontario,
Canada M4P 2Y3 (a division of Pearson Canada Inc.)
Penguin Ireland, 25 St Stephen's Green, Dublin 2, Ireland (a division of Penguin Books Ltd)
Penguin Group (Australia), 707 Collins Street, Melbourne, Victoria 3008,
Australia (a division of Pearson Australia Group Pty Ltd)
Penguin Books India Pvt Ltd, 11 Community Centre,
Panchsheel Park, New Delhi – 110 017, India
Penguin Group (NZ), 67 Apollo Drive, Rosedale, Auckland 0632, New Zealand
(a division of Pearson New Zealand Ltd)
Penguin Books (South Africa) (Pty) Ltd, Block D, Rosebank Office Park,
181 Jan Smuts Avenue, Parktown North, Gauteng 2193, South Africa

Penguin Books Ltd, Registered Offices: 80 Strand, London WC2R ORL, England

www.penguin.com

First published 2014
002

Typeset by Palimpsest Book Production Ltd, Falkirk, Stirlingshire
Printed in Great Britain by Clays Ltd, St Ives plc

ISBN: 978-1-846-14189-8

www.greenpenguin.co.uk

To Hannah, with love

Contents

List of Illustrations

Acknowledgements

I am deeply grateful to Penguin Books for taking this project on and seeing it through, in particular to Laura Stickney, Stuart Proffitt, Richard Duguid and Shan Vahidy. Thanks also very much indeed to Patrick Walsh, my enthusiastic agent, and to all at Conville & Walsh.

A particular debt of gratitude is owed to the Royal Society of Literature and the Jerwood Foundation, whose generous award – so transformative for a first book of this kind – funded research into the history of the Koran and the life of Martin Luther. Thanks also to award judges Robert Macfarlane, Claire Armistead and Tristram Hunt for choosing *The Paper Trail*.

A few formative influences must be mentioned, for both their own magnetic enthusiasms and their generosity: Pamela Clayton, Tom Sutherland and, especially, Barnaby Rogerson. Thanks also to Leyla Moghadam for her crucial help when the idea for the book first took shape. I am indebted to a number of 'China gurus' from over the years, among them Rob Gifford, Yang Xin, David Bray and the encyclopaedic Jonathan Fenby.

My parents have been remarkably supportive and giving throughout the project.

A number of academics, librarians and independent scholars have ben exceptionally generous with their time and wisdom: Desmond Durkin-Meisterernst of the Free University in Berlin on Manichaeism, Michael Marx and the scholars of Corpus Coranicum on the Koran, Graham Hutt of the British Library on China, David Morgan on the Mongols, Gary Williams and Kirsten Birkett on the Reformation, Elizabeth Eisenstein on printing, Robin Cormack on Byzantine history, Lilla Russell-Smith of the Staatlicher Museum, Berlin, on Manichaeism, the Museum of Fine Arts in Boston, and the remarkable Tsuen Hsuin-Tsien on the history of writing and texts in China. I should also mention a few authors on whose works I have been especially reliant: Tsuen Hsuin-Tsien, Michael Suarez and Henry Woodhuysen (for their excellent *Companion*), T. H. Barrett, Jonathan

Bloom, Mark Edwards and Robert Darnton. And thanks to Jeff Edwards for his care over producing such apt maps.

I am very grateful indeed for the three highly accomplished academics who agreed to proofread sections of the text: Edward Shaughnessy, Emran el-Badawi and Andrew Pettegree – their careful and expert comments have made a great difference. There are many friends to thank also, some of them for proofreading: Julien Barnes-Dacey, Nick Watson, Barry Cole and Barnaby Rogerson. Others have been generous with beds at various key junctures – Greg and Audrey Tugendhat, Jamie and Susie Child, Rob and Nancy Gifford, George and Fiona Greenwood, Ali Kille and Rebecca Pasquali, Jes Nielsen and Alex McKinnon. Long-suffering fellow travellers have also been excellent (and patient) company as I researched – Chris Perceval, Julien Barnes-Dacey, Phil Kay and Max Harmel.

As the above shows, I have needed plenty of help, but the most committed, frank, and self-sacrificial support has come from Hannah, my wonderful wife. Thank you so very much. SDG.

I

Tracing Paper

The city of Cambaluc has such a multitude of houses, and such a vast population inside the walls and outside, that it seems quite past all possibility.

Marco Polo, The Travels *(trans. Sir Henry Yule)*

In 1275 Marco Polo arrived at the capital city of the most expansive and unlikely empire the world had yet seen. In the stories he dictated on his homecoming, Marco called it Khanbaliq (or Cambaluc, as in the translation above), the city of the khans, but sixty years earlier it had been a Chinese city with a Chinese name, before the Mongols besieged the city and razed it to the ground. When the Mongols rebuilt it as Khanbaliq, it was just one city among several in the expanding kingdom of Greater Mongolia. But by the time the Polos arrived it had become the capital of an empire that spanned much of Eurasia, from Korea to Eastern Europe. Today it is called Beijing.

For several pages, Marco's travelogue professes his amazement at the scale and splendour of Khanbaliq. He wrote that the four innermost walls of the palace complex were each a mile long, and the four outer walls each ran to eight miles. Eight palaces placed around the inner walls served as arsenals, while a further eight sat between the inner and middle walls. Among them was the khan's own palace, fenced in by a marble wall. The rooms of the palace were covered with gold and silver, and decorated with gilt images of dragons and birds. Six thousand men could sit for dinner in its main hall.

Marco, as a merchant and a Venetian, was well acquainted with desirable objects, and yet even he was astonished by the luxury and grandeur of Khanbaliq. Numbers pepper his description of the Mongol capital, from the twenty-four-mile circumference of the city wall to its sixteen palaces and twelve gates, each gate manned by a thousand guards. Khanbaliq, he recorded, buzzed with commerce and was 'laid out in squares like a chessboard with such masterful precision that words cannot do it justice'. Marco wrote that the city had 20,000 prostitutes and that more than 1,000 cartloads of silk entered it each day. On New Year's Day, the khan received gifts of more than 100,000 white horses and held a procession of 5,000 elephants. Polo's numbers are the stuff of travellers' tales, but the mark Khanbaliq made on him is unmistakable.

Yet something far less magnificent than the city's palaces also caught his attention: the Royal Mint. His diary records what he discovered:

> . . . you might say he [the emperor] has mastered the art of alchemy . . .
>
> You must know that he has money made for him by the following process, out of the bark of trees – to be precise, from mulberry trees (the same whose leaves furnish food for silk-worms). The fine bast between the bark and the wood of the tree is stripped off. Then it is crumbled and pounded and flattened out with the aid of glue into sheets like sheets of cotton paper, which are all black. When made, they are cut up into rectangles of various sizes, longer than they are broad . . . all these papers are sealed with the seal of the Great Khan. The procedure of issues is as formal and as authoritative as if they were made of pure gold or silver . . .
>
> Of this money the Khan has such a quantity made that with it he could buy all the treasure in the world . . . with these pieces of paper they can buy anything and pay for anything.[1]

It is an imperfect and inexact description of Chinese papermaking and Chinese printing on paper, but it quickly became the best-known account across pre-Renaissance Europe. Within the first twenty years of the book's publication, fresh editions appeared in at least five different languages. All this took place within Polo's lifetime, a remarkable success in a Europe not yet familiar with printed books; each fresh edition of *The Travels* was copied out by hand.

Here was Europe's most famous visitor to Beijing marvelling at the

Chinese papers of the Mongol empire. Just as paper was pioneered in China, so too was paper money. By the end of the tenth century, several centuries before his visit, China's paper money in circulation had already reached the equivalent of 1.13 million *tiao* – a *tiao* was a string of 1,000 coins; thus China's currency in circulation could be counted in the millions before papermaking had even reached Christian Europe. Under the Yuan dynasty, in power when Marco made his visit, that number increased substantially. The result of this glut of paper would be hyperinflation.

Yet the earliest mention of significant papermaking even taking place on the Italian peninsula was made in 1276, a year or so after Marco's arrival in Beijing.[2] Before that, Europe's only paper mills were in Muslim-ruled Iberia. Even to a prosperous Venetian merchant, thirteenth-century Beijing and its wares were uniquely ingenious and glamorous.

The paper money of China's Yuan dynasty, produced by its Board of Revenue and Rights, was used as far afield as Myanmar, Thailand and Vietnam (as we now call them). It crossed social and economic barriers too: notes came in twelve denominations, from ten to two thousand. One surviving Great Yuan Treasure Note, as it calls itself on its obverse side, warns that anyone caught counterfeiting the note will be decapitated while the informer will receive a reward of 200 *taels* of silver, equal to perhaps 17½ pounds.

The Mongols had learnt to govern by paper. Illiterate until just a few decades before Marco arrived, the invaders had chosen Khanbaliq as their capital and quickly established an enormous bureaucracy across what is now Tiananmen Square, one that filled three square miles and employed some 10,000 (largely Chinese) artisans. They produced seals, scrolls, brushes, ink, stones and paper – the machinery of the second largest empire in history. (Only the British empire, seven centuries later, was larger.) The medium of this empire was paper, from money to diplomatic letters, from official histories to property records and from palace accounts to imperial decrees.

This is the story of how that soft and supple substance became the vehicle of history and the conduit for landmark innovations and mass movements across the world. For two millennia, paper has allowed policies, ideas, religions, propaganda and philosophies to spread as nothing else. It was the currency of ideas in the most important civilizations of the day, enabling not merely their easy dissemination within their own cultures, but also their absorption by foreign cultures. This role was key to paper's future; just as money, which had been made of clay or metal for millennia, allowed the transfer and exchange of goods and services, so it was that paper fuelled the trade in ideas and beliefs.

What began in Han China 2,000 years ago reached new heights in the Tang dynasty during the eighth century, just as it was spreading into the Islamic Caliphate with its imperial capital at Baghdad, a hub of scientific research and artistic outpourings, before eventually passing into Europe, where it became the tool of the continent's own Renaissance and Reformation. From its East Asian beginnings, this smooth surface rose to become the writing and printing medium of the modern world.

Paper enabled writers to reach an unprecedented number and diversity of readers. Among those writers were the philosophers who, prior to the age of paper, addressed the political vacuum China suffered in the centuries leading up the founding of the Han dynasty in 206 BC. Also among them were the Buddhist translators eager to carry their religion from India and Central Asia into China in the second and third centuries AD, not just to princes and scholars but to merchants, the poor and even women. Among them were the bureaucrats of the Abbasid Caliphate, Islam's second empire, which stretched from Central Asia to the Maghrib, the theologians who used the Koran to help forge an imperial identity, and the scientists and artists which the Caliphate spawned after the newly built city of Baghdad was appointed imperial capital in the late eighth century. Among them, too, were Desiderius Erasmus, Martin Luther and all the scholars and translators who delivered the Renaissance and Reformation from their desks, taking advantage of cheap Italian paper in their quest to recover, reproduce and translate the great texts of Hebrew, Greek and Roman antiquity. Among their number were the censored French revolutionary thinkers of the eighteenth century, for whom Dutch Protestant printers provided such a useful textual outlet. Among them, too, were the few transcendent personalities of the paper age, figures like Johann Wolfgang von Goethe or Lev

Nikolayevich Tolstoy, whose complete works run to 138 and 100 volumes, respectively.[3] Among them, also, was Vladimir Ilyich Lenin, sitting in his tiny study on Clerkenwell Green in London in 1902 and editing *Iskra* (Spark), the newspaper of the Russian Socialist Democratic Labour Party, before it was smuggled into Russia to help foster a Communist revolution.

In China itself, paper's birthplace, it was a crucial element in the launch of the Cultural Revolution in 1966, as students pasted dozens of 'big character' wall posters to the Peking University notice board, criticizing the university director for his conservatism. Sixty-five thousand posters followed at neighbouring Qinghua University, a number the students could only afford with paper as their means. Tens of thousands followed in cities across the country. Within weeks, a little red book of aphorisms, produced to power the revolutionary wave, had swamped China, such that at least 700 million copies were sold or distributed – and the total number of printed copies has since passed a billion. Mao's *Little Red Book*, as it was known, is one of the four or five most-printed books in history.

In the pre-digital age, paper aided the rise of both universal education and universal suffrage. The emergence of thought communities worldwide, whose members read the same books and articles, was enabled by paper too, whether communities of scientists or of composers, of novelists or philosophers, of engineers or of political activists. Paper delivered the Republic of Letters, which transcended national divisions, helping to forge a brotherhood (and it *was* predominantly men) which held their ideas and reading as a common heritage and bond of kinship. Paper was a writing surface cheap, portable and printable enough that books and pamphlets began to be mass-produced and to travel more widely than ever, released from the expense of parchment and vellum, the rarity of papyrus, and the bulkiness of bones, stones and wood. In its spread of knowledge and of ideas, the emergence of paper fomented a revolution.

Yet, for all paper's rabble-rousing power, and its ability to enrich debate and knowledge, we are, today, slowly stepping out of the paper age or, at least, out of the age in which it has no rivals. Paper has already lost the battle with its virtual rival for everyday correspondence, trivia, encyclopaedias and almanacs, reference information, impersonal networking and personal diaries. It is finding a different kind of accommodation with digital formats over newspapers and magazines, government bureaucracy, advertising, religious and political debate, train, plane and bus tickets, office administration and party invitations. An era is passing.

Leaving the paper age does not mean leaving paper behind; we still use stone, iron and bronze, after all, and paper's uses are too multifarious to allow its extinction. There are very few paperless offices or civil service departments. Although Amazon now sells as many e-books as it does printed books, the switch to e-books has already slowed considerably, and readers still like to browse through and buy books they can hold, put on their shelf, turn the pages of, and own. The sensual experience of reading still exerts its hold on us, as does the desire to represent and display our knowledge, attitudes and passions on our bookshelves.

Yet if paper as a surface for written and printed words has shaped history, it has been as Everyman, not as a lord. Its gradual elevation to a symbol of elegance and luxury spells a new age of grand retirement as the surface of expensive invitations, art prints, glossy business catalogues and books aimed at buyers who prize beautiful covers and the physical pleasures of contact and ownership as well as valuable content. It will continue to work freelance in various specialist roles, but it will no longer work overtime. Instead, it will leave much of the heavy lifting of words and images to the internet, to a thousand different digital devices, and so to Microsoft, Apple, Google and Amazon.

Writing aside, paper remains the warp and woof of everyday life, as a brisk walk through your day can prove. Your bedside lamp glows through a paper cover. The painting in your hallway is printed on paper and set behind a paper mount. There is a roll of paper beside the toilet in your bathroom and a paper label on your bottle of shampoo. In the kitchen, your cereal and juice cartons are made of paper and you read paper post, paid for with paper stamps. Your wallet carries paper money. The billboards you pass on your walk to work carry paper adverts and there are children flying paper kites in the park.

The cups in the factory or office coffee machine are made from paper and your work invariably involves at least a little paper pushing.

At lunchtime you buy a newspaper and a sandwich wrapped in paper. Then you buy a present for a friend and have it gift-wrapped in paper. In the evening, a teenager is handing out paper flyers stacked in a cardboard box (also paper) and the lanterns and menus at the restaurant you choose are all made from paper. After the meal your friend offers you a cigarette. Its casing is paper.

It can be eaten as rice paper or folded up as an aeroplane sick bag. Paper can cover a spot and it can cut a finger. It can form a weight that all too often sends your baggage over the airline's limit or it can be caught by a soft breeze and, in a few moments, lifted hundreds of feet into the air. Children have made handheld aeroplanes out of it to fly across their classrooms and demonstrators have burnt effigies printed out on it. It can last for hundreds of years but it can also disappear with moisture in minutes or be eaten by bookworms in a few days. It can be as mundane and practical as a bus ticket or it can be as prized and expensive as the interface of the world's best-loved paintings.

Paper has reinvented itself into hundreds of forms and uses over the centuries. Yet its ingenuity lies less in its ability to switch from supplying cardboard boxes and toilet rolls to kites and Chinese screens than in its role as the porter of the written and printed word for two millennia. All other roles in paper's drama have been the heroes only of sub-plots. Words have been the unquestioned lead, dominating a story that continues to operate on a monumental scale. It is this story that I set out to tell.

More than 400 million tons of paper and cardboard are produced worldwide each year, the same weight as a million multi-carriage trains or almost a million copies of the largest Egyptian obelisk. (That amount would need billions of trees, although paper manufacturers increasingly use spent paper and leftover wood shavings as well as fresh wood and other materials, such as linen rags, for their raw materials.) Laid out as office paper, those 400 million tons would run to 80 trillion A4 sheets. Papermaking is changing the earth's landscape too, spawning vast mono-cultural forests in Indonesia and Brazil, and threatening biodiversity. Because the production of a ton of paper from virgin pulp requires tens of thousands of gallons of water, it is one of the most ecologically destructive manufacturing processes humans undertake. Water tables can sink in its wake. By recycling a ton of used paper rather than making it fresh, 26,000 gallons of water are saved, according to the Bureau of International Recycling. In the US, pulp and paper manufacturers are ranked as the fourth largest industrial emitters of greenhouse gases, according to the US Information Agency. Paper's success through recent centuries owes much to the low cost of its manufacture, but we now know that the costs have thus far been passed on to the natural world and not the market.

For centuries paper has dominated both packaging and the written word. Packaging has a few more recent competitors, but until the arrival of the television and computer screens, paper rose to ascendancy wherever it set foot. It defeated its early rivals for column inches with ease – most significantly, bamboo, papyrus and parchment. Some societies, notably in much of Europe and India, were slow to adopt it, but here cultural factors also played a part; Europe may have feared what it saw as an Islamic product, while oral culture remained dominant within orthodox Hinduism (although not among India's Buddhists or Jains). Ordinarily, however, paper won converts quickly.

Paper is more convenient to carry and store than bamboo, wood, stone, pottery, bone or turtle shells. It is far more abundant than papyrus, whose plants grow well only in Egypt and Sicily. It is cheaper than parchment or vellum, both of which are made from calfskins. Paper is flexible, light and absorbent, and yet it is also strong. You can even fold it up several times. Its chief ingredient must be plant fibres, but these can come from old rags or linen, from flax, hemp, mulberry bark, ramie nettles or from any one of an alchemist's shopping list of plants and vegetables, from mallow to St John's wort to genista.

In China it was traditionally made from the bark of the mulberry tree or the ramie nettle or from old linen rags and fishnet, in a process whose brilliance lies in its simplicity. It was then used by men and women from a thousand different walks of life on every rung of the social ladder. Thus it became the medium not only of historians, monks, poets and philosophers but also of shopkeepers, fortune-tellers, street hawkers and quack apothecaries. And it still decorates the streets of the city that Marco Polo visited 750 years ago.

There is a beggar kneeling on the pavement on the corner of a quiet lane a few streets from Tiananmen Square, his brown face bordered with tousled clumps of black hair, angular as branches. Characters that tell the story of his misfortunes are written out on the sheet of paper set out before his knees, each of its corners weighed down with a stone. Their jet and curvilinear forms stand strong and stark against their white background, in an otherwise grey cityscape. Even a beggar can afford paper, and perhaps, with some ink or paint and a brush, make a living from it too.

Some way to the north-west of the old city, a retired lecturer, now more than eighty years old, lives in an apartment in an old Soviet-styled tenement block amid the capital's most established universities. Professor Yang Xin, like the beggar, has staked his reputation and income on paper and paint. When he traces out his characters on the page, he draws on breathing exercises and seventy years of practising calligraphy to form his subjects. Inside his studio, shelves along the wall and below the large, central table are weighed down with scrolls, books, brushes and paints. The room is sparsely decorated but marked by industry. Yang traces each character in a single movement, decisive but not mechanical. By the time he lifts the bristles off the page, the initial boldness has faded to a few streaks. This easy expressiveness cannot be achieved on China's much older writing surfaces of bamboo and wood. Paper, on the other hand, is inexpensive, encouraging practice, rough copies and risk-taking. It is also smooth, allowing a calligrapher enormous scope for expressiveness.

One of the characters the professor captures best is *chun*, which means 'spring'. Friends have told Yang that his take on the character resembles a young girl dancing. Yang's version is light, vigorous, elegant even, and different viewers claim to spot different images within it. By turns standardized and diversified over the centuries, Professor Yang's art form is writing beyond writing; you probably cannot recognize the contours of the original character in his representation, but you can see the mood of what he describes. It is the visual equivalent of what James Joyce attempted with sounds in *Finnegans Wake*; few readers could decipher the words Joyce uses but their sounds still resonate with meaning. Likewise, even university-educated Chinese struggle to read the characters of Yang's calligraphy, yet the mood of the characters is often crystal clear. They accept his work as an expression of the language so personalized as to be closer to art than script; such expressiveness has been made possible by the partnership of brush and paper.

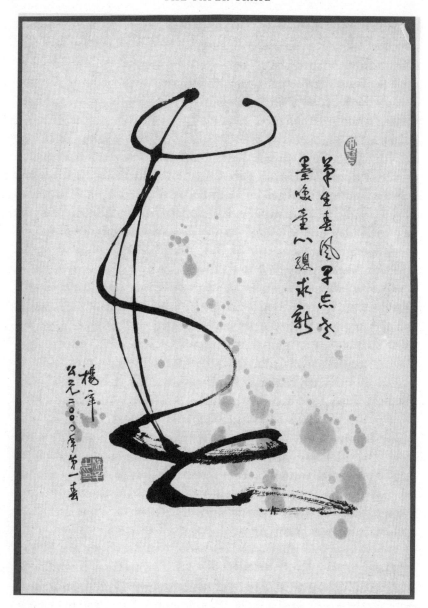

1 Yang Xin's 'Spring' (2004).

Yang's calligraphy is a departure from the pictorial conception of many Chinese characters, a move from bald representation to a more personal, expressive style – Chinese calligraphers have been personalizing their script for many centuries. Moreover, each character is no longer a series of separate strokes but, because Yang's brush is not lifted from the paper, one single and varied stroke. Thus the shape of a piece of writing becomes emotively bound to its creator, allowing calligraphy to convey an individual response while ordinary handwriting is simply aiming to rote-copy. The move towards single-touch characters – characters that can be fully formed without the brush ever leaving the page – began a little over 2,000 years ago and released the script from its traditional confines, risking illegibility but opening up new avenues of written expressiveness.

The beggar and the master calligrapher act like bookends to the many thinkers and workers whose lives or livelihoods have been paper-made in the Chinese capital. Calligraphy was classical China's supreme art form; begging one of its most disdained pastimes. Yet both men made their living by paper, proof of the universality of this Chinese writing surface. Between the social extremes of calligrapher and beggar, the residents of Beijing have, for centuries, used paper for work as for leisure. Fan- and flag-makers, lantern-salesmen, card players and take-away food vendors have plied paper for sport or money. By the sixth century AD, paper kites were used in China for military signalling, measuring distances and for testing the wind. The *Book of Sui*, an official Chinese history written in the sixth century, describes how Gao Yang, a fifth-century northern Chinese emperor, would order a condemned man to be attached to a kite in order to make him fly, thus using paper as a form of execution. On happier days, paper lanterns were used for festivals, notably the annual lantern festival which marks the end of the Chinese New Year celebrations. At court, official historians would make records on paper, while ancient China's enormous bureaucracy ran on paper too.

Missionary translators – Buddhist, Christian, anarchist, communist and capitalist – have all sold their ideas in the city on paper's cheap surface. Paper coincided with the ages of mass religion and mass political ideas too, enabling second- and third-century Buddhist missionaries to China to spread their beliefs and their scriptures more easily. Centuries later, at the end of the nineteenth century, it provided the surface for the newspapers, journals and leaflets which began to undermine the ideological basis of the Qing, the last of China's imperial dynasties.

It became a crucial weapon for the emerging Chinese Communist Party, which grew from a meeting of barely fifty in Shanghai in 1921 to become *the* governing power in China from 1949.

At the other end of the paper hierarchy, road sweepers, ragmen and recyclers have gathered used papers for reuse through the centuries. Recycled paper shops sprang up in the capital, allowing the penny-poor to access its surface too. In a city where the pursuit of universal literacy was only undertaken in earnest during the twentieth century, letter-writers were a common sight on street corners, ready with paper and brush or pen to take dictation from illiterate customers. (Even in 2002, I encountered a merchant letter-writer set up on a street corner in Yining, a city in the far north-west of China.) Beijing, the city where Marco Polo watched paper being made eight centuries ago, was not paper's birthplace, but the Chinese capital has been dominated by paper culture for centuries. It offers a microcosm of what is now a global phenomenon, its beggars and calligraphers just two of the groups whose lives have been shaped by this self-effacing surface.

If you were to follow the routes taken by those papers used by the beggar and calligrapher back to the factory where they were made you could, perhaps, find out who had passed that knowledge of papermaking on to the factory manager too and, in this way, track back through the centuries of expertise and training that has been passed down the generations. You would travel back past the fall of a dynasty and then further back still to its rise. You would pass to a time before Europe even had contact with China and before Beijing was a city and on, further and further back towards the beginnings of Chinese Buddhism. You would follow this genealogy of student-craftsmen until, more than two thousand years ago, probably before even the Qin unification of China in 221 BC, you found yourself watching a craftsman, a gardener, perhaps even a washerwoman, falling upon the discovery of papermaking by accident.

He or she might have left some linen out to dry for too long, or forgotten some rags in a bucket for so long that they disintegrated. Then he might have poured them out over a flat paving stone and ignored them until they had dried, matted together as a sheet, or even just as scraps. All it then took was for someone to paint a character or two onto the sheet with his brush. After all, a literate man might paint onto silk, wood, bamboo or stone, onto the walls of an inn, the surface of a partition screen or on the back of a fan. Even today, many Chinese practise writing characters by painting water onto the pavement; the image evaporates in seconds. Why not paint characters onto a sheet of linen or old rags too?

Or it could have been a more bookish invention, as the great American paper historian Dard Hunter suggested, the brainchild of a resourceful craftsman seeking to reuse the fabric fibres trimmed off the edges of the woven cloth 'books' he was manufacturing. All it required was for these leftover fibres to be matted together like felt before they were used for writing.

This is how you might imagine it. But the official story begins with Cai Lun, the second-century government official traditionally credited with inventing paper in AD 105. Papermaking is in fact at least three centuries older than this, but it was nevertheless Cai who refined paper for more widespread use and who first appreciated the enormous choice of possible ingredients.

I picture Cai standing in his study-workshop at the imperial palace in the 110s, varying his ingredients, experimenting with his pulping and drying techniques, shining his sheets of paper with different stones, testing them out with various inks and paints, toying with colour, texture, thickness, flexibility, toughness and sheen. Then, his patron, the empress Deng, rises up like a conductor and signals the launch of Cai's carefully honed substance across China, in a quest to harmonize the country to this new medium.

And so paper's passage across countries and continents begins. First, courtiers write letters and notes on it, while Confucians begin to brush their theories and politics onto its surface. Imagine it as a flurry of papers spreading across China like autumn leaves through the second and third centuries AD, beginning with Cai and his empress, and gaining momentum as others contribute their own sheets, adding to Deng's modernizing and unifying zeal their own ideas on politics, the cosmos and the family. From the second century, missionary Buddhists start to write on its surface too, filling thousand upon thousand of paper rolls with prayers and sutras, using them to take their message not only to the court and the bureaucrats but to anyone who will read them. Not only are more people now writing; paper has ensured more can afford to buy reading matter for themselves too.

The march of Chinese characters continues into some of the country's furthest corners, as monks, poets and Confucians add their own works to the gathering weight of paper, as the educated write letters and poems, and as the masses buy written sutras as charms or medicine bottles carrying paper labels. As China's neighbours begin to notice this strange phenomenon, so they contract this writing fever themselves. By the sixth century, papermaking is being practised in Korea. Early in the next century, Korean monks introduce the technology to Japan. By now, the old alternative surfaces of wood, shell and bamboo are becoming museum pieces, as China fills with paper.

Paper has already extended its reach through a corridor to the north-west too, filling out into the desert oases of north-west China as Buddhism gains followers and communities. The Muslim conquest of Persia in the mid-seventh century brings Islam to the very borders of China, and in the mid-ninth century the Caliphate adopts paper as the surface of its bureaucracy. By now mullahs, scientists and philosophers are adding their own contributions on theology, politics, art and poetry to the body of papers expanding westwards. Geometry, astronomy, Islamic history and even recipes are written out on paper. Humour and erotica appear too, fusing in the ribald and playful *Arabian Nights*, collated between the ninth and fifteenth centuries, with its graphic sexual metaphors and hyperbolic storylines.

Arab paper begins to wend into Europe too, first Spain and later Italy and Greece. Although Islamic Spain produces paper in the eleventh century, and significant quantities by the twelfth century, the real breakthrough comes in the late thirteenth century, as papers produced in the Catholic states of the Italian peninsula begin to undercut the Arab competition across the Mediterranean, partly thanks to greater access to water.

The late fourteenth century sees papermaking spread north of the Alps and it is here that, beginning in the mid-fifteenth century, copyists are able to put down their pens and set their papers down under presses instead, as movable-type printing produces volley after volley of cheap, thick and printed papers to add to the gathering storm of writings now engulfing Europe, writings fed partly by Roman

Catholicism but especially by the ideas of the Renaissance and Reformation. New forms of writing appear in Europe too, alongside new interests. Giorgio Vasari's *Lives of the Artists*, first published in 1550, chronicles the lives of the great Renaissance artists of Florence in the late thirteenth to mid-sixteenth centuries, pointing to at least two growing interests: the process of art, and secular biography.

Paper has now reached as far east as the islands of Japan and as far west as the far shores of Morocco and Ireland. It has borne on its surface a myriad of hopes, beliefs, discoveries and speculation. It has provided an encyclopaedia of culture and humanity, a great index of ideas and a museum of inked thoughts across the centuries and the continents. Its story does not end there, for in the sixteenth century papermaking is taken from Spain across the Atlantic to Mexico, largely replacing the paper-like *amate*, a Mayan invention used by the Aztecs, and so paper spreads north and south into the Americas, this magical midwife of human words which Native Americans named 'talking leaves'.

This is the great trail that paper blazed across the world. Its journey has, of course, continued, from papermaking's spread through sub-Saharan Africa (following its adoption by Egypt in the tenth century and by the major cities of the Maghrib in the eleventh) to its widespread reception in the Indian subcontinent in the thirteenth century,[4] to its nautical progress on the back of Europe's imperial conquests. But it is paper's journeys across Eurasia and the Maghrib that matter most, since it was these journeys that persuaded the most powerful civilizations of the world to become paper cultures. Much of papermaking's subsequent spread stemmed from the influence and conquests of these same powers.

Paper has been the evangelist of many of the convictions that shaped history, carrying them to distant lands or simply to a ground-swell of people who could never otherwise have absorbed them. Propagandist, tyrant, democratizer, tool, inventor, magician and technician all in one, paper's power lies in its absence of personality. Quietly, inexpensively and often slowly, it has seeped around the world, and history's most galvanizing ideas have hitched a lift on its surface.

Today, there are more paper trails than anyone could count. Each person who reads books, newspapers, office reports, leaflets or labels has formed a new branch of the trail of his own. Every living reader is simply the latest destination in a 2,000-year-old journey that began in China and ends, for the moment, in a billion pairs of hands.

This, then, is the story of how you came to be holding this book, how you came to be following its printed words across dozens of pages made not from bamboo, silk, parchment or papyrus (which you might never have afforded anyway), but from paper. Paper has been your silent tutor, as important to your inner library of information as your parents and your computer screen. You, too, are paper-made.

2

Alpha and Omega

The pages are still blank, but there is a miraculous feeling of the words being there, written in invisible ink and clamouring to become visible.
Vladimir Nabokov, The Art of Literature and Commonsense

Writing was the making of paper. Looking back across two millennia, the early history of paper, before it became the vehicle of the written word, appears like the prelude to a far bigger story. Chinese hemp paper wrappers survive from the late second century BC, vestiges of paper's pre-writing age, while a county map found at Fangmatan, some 200 miles west of the ancient Chinese capital of Chang'an (now called Xi'an), dates from the early second century BC, pointing to a new role. Examples of writing on paper date back to the late first century BC, but they are exceptionally rare and usually little more than scraps.

Marco Polo recorded that the Chinese had used paper for kites (their Chinese name meant 'paper birds'), military signalling, window panes and decorations before there was any concerted attempt to employ it as the surface of the written word. Without writing, paper would still have enjoyed a fruitful, if unspectacular career but, as we look back at the stuttering beginnings of paper's employment, we are like Nabokov at his desk: faced with a blank page but confident that words hidden on its surface are about to become visible. Thus the birth of writing is part of paper's prehistory, the background to the one role it has played which was transformative for the world. And writing is one of our strangest and most ingenious inventions because it preserves what are most fleeting: words.

Words, after all, are how we package our ideas and experiences for each other. As tools of communication they are, of course, flawed; human speech, wrote Gustave Flaubert, 'is like a cracked tin kettle on which we hammer out tunes to make bears dance when we long to move the stars'.[5] And yet those blunt tools remain the most versatile we have; words are still the currency of communication. They salvage our thoughts and experiences from history to live a little longer.

Yet the greatest weakness of words is not their imprecision but their mortality. Often words disappear as quickly as the breath that first carried them. The life of some words can be counted in decades while others are passed on down the generations. But even then they are shovelled into new forms as they pass from man to woman, woman to man. Unchecked, the process of oral history is a game of Chinese whispers writ large.

Yet 5,000 years ago, ingeniously, all that began to change, as words took physical form, sometimes scratched and sometimes carved. Suddenly, they might live on for years, decades and centuries, even if nobody could remember them.

Writing began in Sumer, a kingdom in Mesopotamia (now southern Iraq) that was something like the factory prototype of settled civilizations – it was the site of the first known extensive use of the wheel, irrigation, the plough and the arch. Eight thousand years ago the 'black-headed people', as the Sumerians called themselves, either ceased their wanderings or simply moved south (their origins remain disputed) and built cities, among them Uruk where, in the fourth millennium BC, markings began to preserve speech.

A line drawing of a bison and hunter carved on a wall might tell a story; such images survive from more than 30,000 years ago. But as Sumer's population grew, its rulers needed to keep records of property ownership, trade transactions and temple accounts. The complexity of Sumerian society therefore led to the creation of signs and then, from those signs, to images, images that represented not just objects but the sounds of the names of those objects too.

Gradually adjectives, verbs and conjunctions were added and Sumer drew itself a germinal script. This phonetic genesis, turning copycat images into abstract sound signs, is called the rebus principle. It allows, say, a drawing of an eye to represent the sound 'i' in general, irrespective of meaning; this essential link first took place no later than the thirty-fourth century BC, perhaps in Uruk, the Sumerian capital. The argument over how to define 'writing' remains unresolved, some viewing signs as 'writing' while others insist those signs are only writing once they represent sounds (or phonemes) of some sort. Few would dispute, however, that the switch to a phonetic form was a groundbreaking innovation in the history of communication.[6]

Yet the invention of writing did not mean writing could spread rapidly. Sumerian scribes had to write on clay tablets with a stylus, the writing utensil with which they could carve out wedge-shapes in the clay; 'cuneiform', the name of their script, simply means 'wedge-shaped'. Thus the Sumerian script looks like a set of golf tees set at right angles to each other, even though it had developed from images of everyday objects and shapes, such as the image for a head.

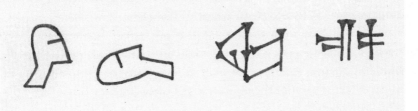

2 Impression of the development of the Sumerian sign 'sag' (head) into cuneiform, from the start of the third to the early second millennium BC.

Tablets were heavy and usually rectangular, though they were sometimes barrel-shaped or even a prism with six or eight sides, each side carrying text. Some tablets were smaller than a credit card, usually those used as receipts for, say, the purchase of a few sheep, a field or a tax payment; these at least were portable. One twenty-first-century BC clay business receipt for the delivery of a lamb (now in the British Museum) measures a little over 1 inch by 1½ inches and weighs around an ounce.

Poems or stories needed far larger tablets; those that recorded significant historical events or political texts were often half-an-inch thick and might measure 12 inches by 18 inches – almost twice as long as an A4 sheet of paper. As a writing surface, clay was at least fairly durable; once the clay had dried, the text was set. But clay tablets did not circulate well, due to their size, weight and fragility. Tiny receipts were manageable but few beyond the court and a few of the elite could afford to own significant texts.

If Sumer could not fully exploit its invention, it was an electrifying technology all the same. Writing changed how we govern, communicate, tell stories, remember important facts, ensure reliability, make ourselves job-worthy, express ourselves and find common ground with different groups in society. For the first cultures to use texts, the invention or discovery of writing was so magical that it needed a divine explanation, as much as the seasons, agriculture, fertility and the sun needed a divine explanation. Plenty were given to explain the miracle, from Scandinavia to Egypt to China.

In Norse mythology (the Germanic Norse script comprised 'runes') the ability to write was earned by Odin, king of the gods, who hung on a tree for days on end by way of payment. W. H. Auden's translation of the myth tracks Odin's journey towards paginated knowledge.

> I took up runes;
> from that tree I fell.

The Kabbalah, handbook of Jewish mystics, views writing as no less touched by divinity. It tells how God chose three consonants (Y, H, W) to form his name, Yahweh, before forging the universe itself from a recipe of twenty-two letters that he later impressed onto the tongue of Abraham. In ancient Egypt it was Thoth, the god of writing, who carried each pharaoh from his death to heaven. In Thoth's *House of Life*, as the empire's largest library was known, scribes composed and copied texts with a pen or scalpel which they called 'the claw that fastens on the tongue'.

Egypt had adopted writing by the late 3200s BC, presumably borrowing the principle from the Sumerians, given their relative proximity. Yet, instead of merely copying the Sumerian wedge-shapes, Egyptians used coloured birds and suns too, a series of erect, thick-rimmed images we call hieroglyphs. The most vivid are filled with blocks of blue, red and yellow. Their columns read like vertical cartoon strips, a series of stick men, knives, ibises, gods, ripples, eyes, ducks, houses, reeds and mountains. Their man-animal hybrids make these tales in stone less picture-book cartoon than *danse macabre*; many were drawn on coffins and the walls of burial chambers. This was a script that covered stone tombs, walls and pyramids, but in the late third millennium BC, it also began to appear on papyrus, that precious reed of which Egypt had all but a monopoly.

The sticky pith of the papyrus leaf was cut into thin strips and a lattice made out of them, ready to be pressed into sheets and then struck repeatedly with a mallet, creating a smooth surface texture ready for writing. Papyrus was the writing material of the Mediterranean for centuries, and thus the writing surface of the classical world – used for its philosophy, mythology, drama and poetry. As the medium for this new civilization, it was also the deliverer of the beginning of the western literary canon. Indeed, the Greek word for the plant's inner sap, *bubloi*, gave birth to the word *biblos*, meaning 'writings on papyrus'. This is the root of our word 'Bible', reflecting the material on which Christian New Testament texts first appeared. Although Papal Bulls appeared on papyrus as late as 1057, papyrus writings largely disappeared from view in the early centuries of the Christian era.

Yet papyrus is not merely part of writing's West Asian and European heritage; it remains, often inappropriately, bound up with its identity to this day. The word 'paper' is itself a misnomer, deriving from the word 'papyrus' even though paper is made through macera-tion (the breaking up and soaking of fibres in water) rather than simply through matting and pounding, the techniques used to prepare papyrus as a writing surface. In Europe, it was long assumed that paper was a Greek or an Arab invention; either way, it was supposed, the secret of papermaking must then have been passed eastwards to China. The words 'paper' and 'papyrus' have often been used inter-changeably, as though their meanings were identical.

This confusion over paper's origins, geography and chronology continues. Even the 2008 English Standard Version (ESV) of the Bible translates one of the final sentences of the apostle John's second letter as 'Though I have much to write to you, I would rather not use *paper and ink*.' All major English translations use the same final three words, from the King James Version of 1611 to the New American Bible of 2011. Yet John, writing in what is now Turkey, would not have been able to use paper for some seven centuries. The ESV gives a more accurate picture of ancient reading culture in Paul's second letter to Timothy, when Paul asks Timothy to 'bring the cloak . . . the books and above all the parchments' (2 Tim. 4:13). Parchments, made from animal hides and in use from the third century BC in west-ern Turkey, were difficult to produce and therefore more expensive, and so writings on them were usually more valuable in content; per-haps in this case they were Hebrew Scriptures, whereas 'the books' Paul refers to were doubtless on papyrus, which was also used for such pap as notes and receipts. As late as AD 390, St Augustine apol-ogized for writing a letter on papyrus instead of on parchment.

By comparison with the clay tablets produced in Sumer 5,000 years ago, papyrus was infinitely cheaper and easier to produce, and far more portable. It became the writing and reading surface of both ancient Greece and ancient Rome. In ancient Greece, it enabled the growth of writing, which was used in banking, in commerce, in citi-zenship documents, in list-making and, at the higher end of society, in philosophy, poetry and politics too. By the fifth century BC, an *orchestra*, or semi-circular recess, in Athens's marketplace – the *agora* – was allocated to booksellers. As democracy expanded in the city from the fifth to fourth centuries BC, literacy became increasingly indispensable for those in government. The Athenian practice of ostracism, by which an individual could be expelled from the city-state and which required eligible citizens to cast a written vote, has allowed modern historians to arrive at a basic literacy rate of 5–10 per cent for Attica in the fifth-to-fourth centuries BC. However, this refers only to a very basic, functional literacy, far removed from the more wide-ranging ability that 'literacy' usually implies today. The Hellenization of Alexander the Great's empire also raised educational provision (and, with it, literacy levels) and installed libraries in various cities around the Eastern Mediterranean, most famously at Alexandria in Egypt.

In ancient Rome books were used not merely in education and high society but in the military and other spheres of ordinary life. Where the Greeks had exported knowledge, the Romans imported it, amassing books from conquered territories to fill libraries at home. As collections grew and copying and translation projects gained momentum, so textual criticism developed too. Bookshops were concentrated in a particular area of Rome, and literary culture aided social mobility as the city became the centre of the Mediterranean book trade. By the end of the first century AD, it was possible to buy codices in Rome as well as scrolls and to select from Greek and Roman titles, old and new.

Yet it was paper, not the papyrus scroll, which would achieve the greater victory for the written word. Although literacy was growing in ancient Greece and Rome, it was too limited for popular reading to flourish, especially as papyrus, though not prohibitively expensive, was not cheap either. Greek texts generally lacked spaces between words – a key exception was in lists – which acted almost as a veto on silent reading and contemporary references make clear the practice was exceptional. In Greece, only the elite could do more than draw up a few lists.

Rome itself undoubtedly had a significant book trade, but evidence for similar book trades in other Roman cities is limited. Claims of textual inaccuracy were widespread and Cicero himself cited limitations on access to books, while many poets besides were sceptical of the value of committing their works to a physical form, despite the publicity it might bring. Whether Augustan Rome was primarily an oral or a book-based culture remains a contentious question. But the ability to read much more than signs, lists or adverts was confined to a small and powerful section of the population; the use of written ballots suggest basic literacy of up to 10 per cent in the Roman empire from the second to fifth centuries AD.[7]

Moreover, in both the Greek city-states and in the Roman empire, publication of a new work began with a public recitation and, if a work was reckoned suitable for the public, it might continue to be delivered out loud – only more private works would be read in silence. Papyrus scrolls also limited the practicability of longer texts, since it took so long to roll forward or back to a different section of the text – usually texts were no longer than 30–35 feet, although that would only allow for two or three of the twenty-four books of *The Iliad*. In both Greece and Rome, even the concept of a widespread population able to read independently (with all the choice that implied) was foreign. Thus, while both cultures placed spaces between non-vocalic words and both cultures would sometimes insert vowels into their writings too, neither culture introduced both together, which would have aided private, silent reading enormously. (In the central Middle Ages, however, word spacing *was* introduced, permitting easy silent reading, which in turn enabled less fluent readers to read without pressure – and so gradually come to understand more difficult texts.[8])

In short, the physical texts of the Greco-Roman world remained closely bound up with recitation, oratory and public entertainment on the one hand while, on the other, their format – the papyrus scroll – hampered easy reference, portability and the reading of longer texts. Papyrus was fragile and thus impossible to fold or bind, and text would appear on only one side of the material. It was also laborious to make it smooth for writing. Storage was not simple either, since papyrus is especially sensitive to moisture and the papyrus plant, although not limited to Egypt, grew most abundantly in the Nile Delta and along its banks.

Papyrus did help to deliver a cultural outpouring in both Greece and Rome. What it could not do, however, was provide a culture of books in which reading and writing were inexpensive, longer books manageable and storage straightforward. Nor could it overturn oral culture. Its place in the history of literacy was assured, but it would be far superseded.

During the second millennium BC, writing caught on among the Elamites in Iran, the Harappans in the Indus Valley (now Pakistan) and the inhabitants of the North China Plain. While the Elamites and Harappans wrote onto clay tablets or stone monuments, writings in second-millennium BC China were on turtle shells – and probably on bamboo and wood strips as well, although it is the shells that have survived. These scratched images usually marked major events such as a royal birth, a battle or a drought; even an untrained eye can sometimes recognize a baby, a tree or a house. Although hardy and not too cumbersome, turtle shells cannot accommodate much text. Besides, they were seemingly used not to pass information to a wider or more distant audience but simply for their supposed magical properties.

As new scripts appeared, they were borrowed and adapted by different cultures. It was probably slaves who borrowed from the Egyptians to forge the beginnings of a Canaanite script in Syria by the start of the second millennium BC. From Canaanite came Phoenician, sometimes called the first alphabet since it employed a different sign for each consonant (although none for vowels), and from Phoenician came Aramaic, Hebrew and still others, until Western Asia was a patchwork quilt of alphabets. As the Phoenicians plied the Mediterranean in their trading ships, new alphabets appeared on islands such as Sicily and Cyprus, as well as in Greece.

Following the conquests of Alexander the Great in the fourth century BC, Greek culture engulfed the Eastern Mediterranean. Its writings spread too, and the Greek alphabet became a settlement for stories, ideas and poetry far from its ancestral home. From Greek, the Etruscan script formed a bridge to the Latin alphabet, its written shapes the book masters of the distant future, following the spread of a Latinized Christianity across Europe.

Thus writing appeared in waves, spreading from its Chinese and Mesopotamian roots across Asia and Europe, the descendants of the

Sumerian and Chinese scripts becoming the lodestones of new identities and literary canons across the continents. Just as Chinese spawned new Korean, Japanese and Vietnamese scripts, so Greek and Latin brought fresh alphabets to Europe, among them the Gothic alphabet, in use in Bulgaria from the fourth century AD before it was transported across large swathes of Europe.

Finally, on the Arabian Peninsula, close to the birthplace of writing, a new script, descended from Aramaic, appeared in the fifth century, and was quickly carried in all directions: east to Persia and the edge of China, south to the ports of Oman and Yemen, west to Egypt and the Maghrib, and north to Turkey and Armenia. It overran all it encountered, the most powerful script yet, its letters swooping across the multicultural pages of these lands like the rhythmical tails, dots and rests of a music score. It had every spur it could hope for – political, economic, cultural and religious – for this was no less than the bright-eyed adolescence of the Arabic script.

Yet, while it had been written out on bones, bark, stone and parchment, Arabic had lacked the surface that could fully deliver on its ambitions.

As this Babel of scripts spread quickly around the world, so cultures of writing were emerging that would deliver to paper its global future. Nevertheless, paper was born into one particular culture: China. Until the second half of the twentieth century, Europeans generally assumed that all scripts could be traced to one ancestor; just as they believed the roots of all world civilizations lay in Mesopotamia, so they assumed that all scripts, even ancient Chinese, must find their roots there too. The problem with this theory of a single global mother script, an *ur-sprache*, was the geographical gap that lay between the scripts of the ancient Middle East on the one hand, notably Sumerian, Akkadian and Egyptian hieroglyphics, and the script of ancient East Asia, namely, Chinese. Some scholars came up with theories to plug the gap, lured by the romance of the pursuit. After all, if the beginning of writing is not the beginning of human history, it is at least the frontispiece of humanity's autobiography.

In fact, writing had at least three ancient and independent beginnings: Sumerian, Chinese and, by the third century BC, Central
American. Even the Egyptians, the Harappans and the Elamites might
have coined their scripts from scratch, although the case is far from
clear-cut. As we have seen, the Sumerians probably first set stylus to
tablet to cope with the complexities created by forsaking nomadism:
to manage an agricultural surplus, protect land ownership rights and
keep on top of tax receipts, for example. In China, on the other hand,
writing became viewed as a way of ordering the world at a much
grander level – if not a second Chinese creation to match the birth of
the earth itself, then at least a means of restoring primal harmony,
part of a quest to return to a golden age. The ingredients of this written power were the practice of magic and the study of history; through
the correct reading of history, it was possible to discern the means of
political and social harmony.

Even the earliest pages of China's prehistory emphasize the Chinese
obsession with writing. By the second century BC, the legend of the
Yellow Emperor, supposed founder of the Han Chinese state in the
third millennium BC, had been set to writing. According to the story
– and there are several versions – the emperor's court historian
Cangjie, a man with four eyes, was hunting one day when he noticed
the network of connected veins on a turtle. Impressed, he began to
study the natural world and to coin a system of symbols for natural
phenomena. These, the ancestors of phonetic characters, were known
as 'oracle bone script'. Other versions of the legend refer to other animals, but the turtle legend is especially striking in light of recent
discoveries.

In 2003, American and Chinese archaeologists were digging up 349
graves in Jiahu, a prehistoric site in central China, when they discovered a group of turtle shells and animal bones. Eleven different images
had been scratched onto the shells and bones 8,500 years before and
some resembled Chinese characters, especially images of an eye and of
the sun. Nine of the eleven images were on the plastron and carapace,
the outer and inner parts of the turtle shell. In one grave, archaeologists found a decapitated corpse (the head was never found) and eight
turtle shells. The characters were not phonetic; in other words, the
markings were not complete writing. The rudiments of a Chinese

script may have developed on other surfaces too, but turtle shells and animal bones are the earliest Chinese writing surfaces to survive.

True writing in China was born several centuries later. Moreover, plenty of examples were uncovered in the nineteenth century. Yet instead of being passed on to museums or research institutes, they were used as medicine. This revelation emerged thanks to a medicine purchase made by a Beijing philologist, Wang Yirong, in 1899. Inside the parcel Wang had bought were bone fragments that were meant to be ground up as 'dragon bones', a salve for cuts and bruises. (Turtle shell fragments were also ground up and used to treat malaria.) Wang noticed scratched images on the fragments. More importantly, he found that he could read some of them. Wang and a fellow researcher, Liu E, used their knowledge of ancient Chinese bronze inscriptions to decipher the scratches. Although Wang himself committed suicide in 1900, after an international force had occupied Beijing, his colleague Liu continued to buy up all the shell and bone fragments he could find in the medicine shops of Beijing, and in 1903 he published *Tieyun Cang Gui* (literally 'Iron Cloud Tibet Turtles'), a book of 1,058 rubbings of shell and bone inscriptions he had collected. Liu and Wang's discoveries pointed to how the Chinese script may have been born, namely, through the practice of divination.

The second-millennium BC *Book of Changes*, one of China's oldest texts, is a divination manual. It explains the eight trigrams – symbols that stand for heaven, earth, fire, water, air, thunder, mountain and lake. Apparently random routines (like blindly grasping a handful of yarrow sticks from a larger bunch) would point the practitioner to a particular number. He would then look up the number in the divination manual to read the linked trigram; the trigram would indicate what course of action he should take. Confucius, living in the sixth century BC, read his bamboo *Book of Changes* so assiduously that the binding fell apart three times.

Wisdom-seekers could also tell the future by boring small hollows on the back of a turtle shell and inserting hot brands into them; each hollow would send a single crack out across the surface, which the diviners would interpret. In some shells that were discovered at a site at Anyang dating to 1200 BC (the same site that Wang's own quarry was traced to), the question, the divined answer and the outcome had all been written down.

> Three weeks and a day later, on *jiayin* [day 51 of the ritual cycle], the child appeared. Unlucky. It was a girl.

The questions were formulaic but the answers were often expressive; the Anyang dig had yielded true writing. The written lexicon at Anyang included 3,000 characters, and one fragment even mentions a scribe. After all, when writing appeared in China, signs had already been used for several thousand years. No language is perfectly phonetic; English, for example, has a small arsenal of signs, such as £, %, &, and 0 to 9. But truly phonetic writing *can* represent the sounds of a language as well as the ideas, and the Shang dynasty scratches at Anyang do just that.

Anyang shows the Shang honouring ancestors, reading the future and placating the gods. But the Anyang inscriptions are also an inventory of human desires, of their fulfilment and their denial. True writing in China came of age here, in characters like line-drawings, their forms dictated by the brittle surface on which they were etched. And almost nothing has proved to be as enduringly Chinese as its characters. Other markers of Chinese identity and culture – Confucius and the classics, the imperial system, polygamy, the system of palace eunuchs, Buddhism, and even the authority of parents and teachers – were all either outlawed or attacked in China over the course of the twentieth century. Characters, on the other hand, although partially simplified under Mao Zedong, have proved to be the cement of Chinese civilization, providing a link between the ages, even into communist times; indeed, in 2010 there was an exhibition of Mao's own calligraphy in his home province of Hunan in southern China. It is hard to conceive of China remaining itself without its script.

Writing in China was the arbiter of both the future and the past, the tool of both diviners and historians. While other ancient cultures used writing, China revered it, turning it into an agent of magic, an interpreter of history and a proof of authority. A new age of writing emerged in China as a result, setting it at the heart of the culture. Three millennia ago, there was nowhere on earth writing had risen to such heights.

3

Preparing the Soil

*Writing is the great work in the management of the state; it is
an accomplishment that does not expire. A life span ends after
its allotted span; and with the death of the body, honour and
glory pass away too. Life and body are limited by time, unlike
Writing, which is eternal. Therefore the writers gave over their
bodies to ink and brush, and materialized their thoughts in
tablets and collections. Without the praise of historians or the
patronage of the powerful, their names and their reputations
survived for posterity.*

Cao Pi, Essay on Writing, *third century* AD

Writing could almost be classed an inevitability. Sumer, China and
Mesoamerica all developed writing unaided, giving the practice, as we
have seen, at least three birthplaces. Writing's early uses spanned such
different practices as fortune-telling in China and accounting in
Sumer. This breadth of subjects, from magic to measurements, the
future to the past, reflects the universality of writing. Its birth needed
a sufficiently complex and powerful civilization in which the spoken
word was no longer enough.

Paper, on the other hand, was a child of chance. By the middle of
the eleventh century BC, when China's Shang dynasty was overthrown
in battle, writings in China had begun to expand beyond magic. As a
result, their surfaces would become more important; bones, shells and
stones were still used, but only bamboo would prove equal to the
demands of the age. Centuries later, it was a bamboo-writing culture
into which paper would be born. The invention of paper cannot be

attributed merely to necessity as the invention of writing surely must be. The Chinese might have kept writing on bamboo for another millennium, had paper not appeared. Yet once it did, paper's superiority as the courier of words became plain.

However, convenience alone could not propel paper to become writing's leading partner. Bamboo was inseparable in China from the high and ancient wisdom it carried on its surface; long before the emergence of paper, Confucius himself had written on bamboo. It was a treasured surface for words, lending respectability to the content. In its earliest days as a writing surface, paper was seen as far inferior in quality to bamboo, despite its practical advantages; yet it was bamboo bibliophilia which would create a world of writing for paper to fuel. Political turmoil, beginning in the eighth century BC, would spark a search not simply for military might or signs in the sky, but for fresh ideas about the cosmos, government, social relations and the basis for peace. This war of words was played out on bamboo.

Three dynasties in particular turned writing into the currency of power and governance in China: the Zhou (1046–256 BC), the Qin (221–207 BC) and the Han (206 BC–AD 220). The later rise of paper only makes sense in light of the culture of writing that had first emerged, on bamboo. In *Red Dust*, his travelogue on post-Mao China, the author Ma Jian wrote that China's roots went too deep and that he was dirty from all the digging. That is the danger of heading far back into China's prehistory, but paper's story cannot be told without a little digging.

In the ancient Shang dynasty, during the late second millennium BC, writings in what is now China appeared not only on turtle shells, cattle scapulae and other animal bones, but also on the surface of expensive bronze tripods, used in religious rituals to make offerings to the gods. With its capital at Anyang on the North China Plain, the full extent of Shang control is difficult to gauge, but its influence was felt as far as today's Beijing region, hundreds of miles to the north-east. The Shang dynasty used writing for symbolic as well as magical power.

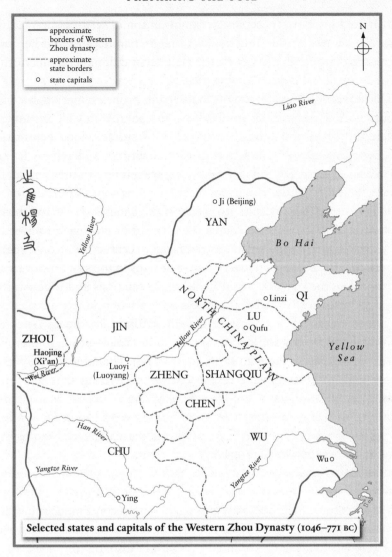

Selected states and capitals of the Western Zhou Dynasty (1046–771 BC)

3 The Western Zhou dynasty (1046–771 BC) comprised an alliance of many states under the official aegis of the Zhou, with its capital at Haojing. Only some of these appear in the map above, several of them the geographical outliers. It is from this period that the earliest extant reference to *Zhongguo* ('central kingdom') survives. Today the word is simply translated as 'China'; it may originally refer to the position of the Zhou on the plains, or even to the Zhou's sense of civilizational supremacy. Through its centre flows the Yellow River, known both as the cradle of Chinese civilization and as one of the most destructive rivers in human history.

The Zhou dynasty, however, would be far more important for the spread of writing in China, thanks in part to a much greater geographical reach. Beginning in the eleventh century BC, it ruled from the dynastic capital at Haojing near the Yellow River over an empire with a clear political hierarchy and a strong military. The Zhou would last twice as long as its predecessor, the Shang; as well as ruling a larger territory, the dynasty would also introduce far more advanced forms of government and bureaucracy, laying the foundation for the imperial system. The Zhou was the *primus inter pares* of a confederation of states which it allowed relative independence in return for assurances of loyalty and regular tribute. Each state operated as a royal fiefdom of its own and treaties between them were formalized by attaching bamboo strips of writing to an animal, which was later buried. The Zhou managed to oversee this network of states by expanding the bureaucracy to forge the beginnings of Chinese imperial governance. It filled the bureaucracy with scholar-mandarins, men conversant not only with governance and diplomacy but with history and poetry too. Zhou poetry celebrated martial virtues and ceremony, reflecting a way of life centred on political order and ritual. Its poems embraced the agricultural seasons as well as the mythological founding fathers of the Han people. The Zhou introduced iron, bronze smelting and irrigation to increase harvests, and the dynasty kept back a ninth of all arable produce in reserve. Peace came with a surfeit of weapons and crops.

The Zhou rulers were atavistic, pickling their history in bamboo archives and on the surface of ornamental bronzes in order to confirm and perpetuate their political legitimacy. Some 6,000 inscribed bronzes survive from the Zhou dynasty attesting to the enormous influence bronze writings have exterted on Chinese textual culture. The best of these bronzes were used for ancestral sacrifices and would have heroic deeds recorded on them. The writing was much better ordered than on many of the divination bones and set out in columns, with the first column sitting on the right (although some of the bone-writings used columns too). Some inscriptions related dynastic or mythological strories at considerable length, others acted as legal documents, while more than seventy Zhou bronzes that survive are inscribed with poetry. A few bronzes use ornamental scripts, notably 'bird script'.

Bronze may have been the first surface on which Chinese scribes employed artistic writing styles.

Daily business at court required much larger quantities of writing, however. In the eleventh century BC, a Zhou scribe would write out the king's will and testament and copies of state documents were filed in the archives. The early Zhou had a Grand Recorder, an Assistant Recorder, a Recorder of the Interior and a Recorder in Attendance on the Sovereign. Writing was the royal preservative. This written unity would knit together its polyglot empire, as the Zhou rolled out a more standardized script. That script became the glue of Chinese civilization, the tool of mandarin-scholars across the North China Plain.

Although writing remained largely an activity for the elite, under the Western Zhou (1046–771 BC) the elite found new uses for it, such as the ancient form of 'letters of appointment' sometimes referred to on Zhou bronzes, or mention of the king handing a document to a document-maker. The use of both bureaucratic and ritual writings points to an increasingly text-savvy elite. Indeed, there even seems to have been written administration practised at local level: trade transactions, legal cases, military administration and land concessions all involved basic written records.[9] The Chinese word for 'book', *tushu*, literally means 'map document'; it emerged in the Zhou dynasty.

The glacial decline of the Zhou dynasty took half a millennium, beginning with the enforced move of the capital city, after a siege, in 771 BC. Its position as patron of the peace then slowly crumbled. As the possibility of dynastic extinction became increasingly tangible, Zhou rulers made plans to save their libraries as well as their lives. In 517 BC the king's son was forced to flee the capital with a retinue of his staff. Yet such was the importance of owning the history books that the fugitives carried the Zhou archives with them. Just as Zhou ascendancy had embedded reading and writing at the centre of Chinese civilization during the Western Zhou period, so now the chaos of dynastic decline during the Eastern Zhou period (771–256 BC) sparked an age of inquiry that raised the role of authors far higher than the Zhou rulers had intended.

*

Books under the Zhou looked very unlike the books we buy today; instead of broad pages their books comprised narrow bamboo strips, and where we use a sheaf bound on one side they used a roll-out mat joined by strings. Nevertheless, they were still books, since they bound pages together to form a continuous whole. Bamboo was cheaper, more versatile and easier to write on than the alternatives of the day (notably bones, bronze or stone). Moreover, bones and bronzes had specific and limited uses. Bamboo, on the other hand, would prove to be an apt partner for a more bookish age. It lasted as a writing surface from the eleventh century BC to the fourth century AD, capturing on its surface a millennium-and-a-half of Chinese civilization.

Two figures from this period dominate the view, from any angle. Centuries later it is easier to admire their impact and stature than it is to examine the details of their lives and personalities, which are scant and muddied by mythology. We can only get so close to Laozi ('Old Master') and Confucius ('Master Kong'); indeed, there is debate over whether the Laozi of legend ever existed. According to Sima Qian, the great second-century BC father of Chinese historians, Laozi was an expert in history at the royal archives of the Zhou in the sixth century. When he met Confucius, Laozi told him to discard his airs and master his lust, while Confucius told his followers that Laozi was an enigma, like a dragon miraculously surfing the clouds to heaven.

Yet for all their mystery, these two figures were the trailblazers of a new kind of knowledge. Neither sought revolution; indeed, Confucius spent many years trying to find an *existing* ruler who would adopt his social and political philosophy. Yet both men wrote out their ideas without the usual political patronage and outside the institutions which had traditionally controlled the historical record. History writing in the kingdoms of ancient China had been owned by its rulers, and the recorders of history had worked as court scribes, propaganda secretaries for the ruling houses since the early Zhou dynasty. But Laozi and Confucius, however unwittingly, loosened the bindings that tied writers to their ruler, signalling new power for the author. This new power would help to forge the educated Chinese man of the future, too.

A Chinese proverb, which has since passed into cliché, described a Chinese gentleman as a Confucian in public and a Daoist in private. (It was sometimes added that he was a Buddhist on his deathbed.) It is a saying that betrays where China was heading, namely, towards becoming an empire of writing and of books, particularly the writings of Daoism, Confucianism and, later, Buddhism. Writing-based government bureaucracy was already well established in China, but Laozi and Confucius stand at the beginning of a new Chinese age of the book.

According to the official imperial histories, Laozi first wrote his thoughts down some two and a half millennia ago, on a broad hill west of Xi'an. It is now home to the memorial temple of Youguan which lies, low-slung, across its brow. Within view stand the Qinling Mountains. Bamboo forests and peach trees pepper the nearby hills and fields of maize lie scattered across the valley below. These mountains were the favoured hideaway for troubled courtiers, hermits, poets and wanderers through much of Chinese history.

4 Daoist monks at Youguan Temple in the foothills of the Qinling Mountains in 2009. The site of the temple is traditionally believed to be the site where Laozi wrote down his *Five Thousand Words*, the *Dao De Jing*. In 2009 the oldest monk at the monastery, seventy years old if he was a day, lived in privileged seclusion close to the summit of a nearby hill.

Living in the sixth century BC, Laozi was more pessimistic than most of his peers about the Zhou dynasty's prospects. As the state looked to liturgies and ritual rubrics to restore the dynasty's halcyon years, Laozi lamented that merely imposing rules and restrictions on men was proof that the world had lost its sap of goodness. Finally, wrote the historian Sima Qian, Laozi harnessed his ox and journeyed across China, leaving the capital for the mountains, fifty miles south-west of the city. Here he came to the Western Gate which led out of China, out of settled civilization and beyond which he might find the great paradise in the West, which may simply be a euphemism for heaven itself. To leave the kingdom, after all, was to forsake life. The gatekeeper recognized him at once and cut him a deal: if Laozi would write down his wisdom, the gatekeeper would open the door for him. The result was Laozi's 5,000-word work, the *Scripture of the Way and Virtue*, or *Dao De Jing* in Chinese.

> When the way was lost virtue remained
> When virtue was lost benevolence remained
> When benevolence was lost righteousness remained
> When righteousness was lost there were the rites.
> The rites are the wearing thin of loyalty and trust,
>   and the beginning of chaos.

Laozi's *Way and Virtue* forms a collection of aphorisms and prov-erbs whose aim is to unite the disciple with the natural order of the world, not through knowledge alone but through intuition and experi-ence. They enable a disciple to discover harmony with nature and to harness it to his or her own needs. After the Bible, the *Dao De Jing* has had more translations worldwide than any other book, despite the endless debates over its author's identity and its true vintage.[10] The *dao* (or *tao*) of Daoism means 'the way' and Laozi believed he had uncov-ered it, although his descriptions of how to find it are often as cryptic as they are illuminating. Daoism has operated as China's antidote to urban life for centuries, offering an escape from the strictures and pat-terns of court life. Yet that same ritual Laozi wished to escape was very dear to the followers of Confucius, and only Confucius rivals Laozi in his influence on imperial Chinese thought and writing.

Confucius was six foot tall, probably divorced and often contrarian.

He had a humanist's faith in human nature and a moralist's disappointment with the rulers of his day. He bore the hallmarks of a social conservative in his love for loyalty, the ancients, propriety, public humility and an inflexible social structure. He preached a utopia in which each man knew his place and stayed there, a utopia plucked from the borderlands of myth and history. It was for such reasons as these that in 1966, as the Cultural Revolution attacks on anything old-fashioned gathered steam, students dug him up to prove him (and his ideas) dead.

But his politics was no mere throwback. Confucius believed governors should be chosen for their virtue and ability, not their lineage, and that the point of good government was the people's welfare. He taught that a ruler who was not righteous and humane would forfeit the Mandate of Heaven – his divine right to rule – because the ruler existed for the sake of the people he ruled.

As Minister for Crime in the state of Lu, Confucius introduced softer punishments for minor crimes, abolishing the practice of cutting off a person's nose for petty theft. But in 498 BC he left his home and his job, ostensibly in pique at the king's failure to observe ritual. He wandered around eastern China for fourteen years with a group of disciples. He was outraged at the moral decline of the Zhou, and believed only a supreme emperor could restore ethics to the kingdoms; morality and political unity were natural bedfellows. Yet he saw only destruction on the horizon.

Confucius is traditionally credited with *The Analects*, one of classical China's *Four Books*, and *The Spring and Autumn Annals*, one of its *Five Classics*. Of all the nine titles, *The Spring and Autumn Annals* was considered the supreme work, because it alone was reckoned to be able to war with evil, outline the way of heaven and rectify right and wrong. The book was the founding work of his belief system, the ur-text of Confucianism. The era itself would come to be named after this one work.

For all their influence, the annals are rice writing; they lack colour and texture. Without descriptive narrative, they chronicle a 242-year procession of state visits, floods, temple rituals, marriages, military victories, city wall construction projects, regicides, even plagues of locusts. Their bald prose uses just 16,771 characters to cycle through two-and-a-half centuries. The shortest entry, for the year 715 BC, is one word:

Pests

Later scholars, however, found the *Annals* so gnomic and ingenious that arguments over interpretations spiralled into colloquies until five commentaries dominated. One such commentary was eleven times the length of the *Annals* itself. But all the commentaries agreed that the *Spring and Autumn Annals* was an inventory of the failures of the Zhou dynasty. On their overall diagnosis of the dynasty, therefore, Confucius and Laozi were in agreement. An archaeological find made in northern China in 1993 suggests they may have agreed on much else besides.

Dubbed 'China's Dead Sea Scrolls' by one of the researchers working on the project, the 13,000 characters set down on 804 bamboo strips included a jumbled *Dao De Jing* and several Confucian works. They point not only to a rich culture of religion and writing, but also to overlaps between Confucianism and Daoism, perhaps even to some kind of conversation between the two. The writings dated to the fourth century BC, but the conversation would continue. Confucius and Laozi's heirs would brush out new ideas on politics, philosophy, relationships and religion, turning bamboo strips into the leading forum for debating ideas across the empire.

A great shift now took place: instead of literate subjects simply looking to the state to fashion history, a new authority emerged, one based on texts and on the interpretation of those texts. Historians would no longer simply be state servants; instead, their clout and independence began to grow together. Confucius had, one of his greatest heirs wrote, signalled reformation by writing his *Spring and Autumn Annals* without a patron, and so challenged the right of the state to control history. By the 320s BC, Confucius's ideas were being discussed in a scholars' academy in Linzi, one of the state capitals, as mandarins and exegetes argued over whether humanity's self-reflective, self-critical faculty proved its goodness or whether morality was in fact counter to human nature. In the process, both the cosmos and the state itself were put under the microscope.

As political breakdown in China continued, so the old ethical debates, which had sought to always work from first principles, were replaced with the study of political realities, signalling the ascendancy of a newly pragmatic and empirical approach to politics. (A comparable shift would take place through the 1910s and 1920s, as the founders of the Chinese Communist Party in Shanghai abandoned the high ideals of anarchism for an activist's communism.) The Linzi academy itself disappeared in the mid-third century BC, a victim of shifting political priorities, but it had helped to raise the written word to the status of an authority in its own right, the page becoming a battlefield for ideas and a place to argue over the direction of politics in the future, usually through the process of interpreting the classics.

The written apotheosis of these debates came with the publication, in the mid-third century, of an encyclopaedia under the chancellor of the state of Qin in central China. The official histories recount that the chancellor invited 3,000 scholars to his home to help with the project. He was reported to have 'exhausted all the writing strips, brushes and ink' to pen a syncretistic work that argued the empire existed for the benefit of the ruled, not the ruler. Its ideas encoded China's 'Hundred Schools' period, which lasted from the fifth to the third century BC. The period's very name bears testament to the innumerable new philosophies it bred, all brushed out on bamboo.

Just twenty years after the chancellor's death, however, all his work was undone. In 221 BC the state he had served conquered its neighbouring states and beyond, uniting China into a true empire and insisting on the doctrine of Legalism. (Legalism argued for a law-based system of governance but lacked any ideal of public service.) Now a vandal power, this Legalist force despised Confucius, believed the state existed for its ruler, not for the ruled and set about the greatest project of centralization yet, standardizing China's characters, weights and measures and bureaucracy. Its name was the Qin dynasty.

The Qin's new chancellor, Li Si, has gone down in history as the greatest vandal of all. According to the historian Sima Qian, Li advised the first emperor to burn all the books in the imperial bureau and to tattoo any hoarders, the result in part of a Legalist philosophy that valued hierarchy and order over learning, particularly Confucian learning. (Criminals and bandits had been regular recipients of tattoos since the Zhou dynasty.) Anyone who discussed the old Confucian classics the *Book of Documents* or the *Book of Odes* should suffer execution and his corpse should be put on show by way of warning. Anyone who used the past to criticize the present should be killed with his family. A year later, Sima Qian recounted, the emperor buried 460 scholars alive. Yet for all its success in centralizing the empire's governance, the Qin failed to turn China away from its books; indeed, some of the dynasty's more bookish opponents grew legendary.

Fu Sheng was a scholar of the Confucian classics and the words of the *Book of Documents* were his staple. As the campaign to burn the books and kill the scholars took hold, Fu blinded himself, feigned madness and probably fled from home. When the old learning was revived under the Qin's successor dynasty, the Han, Emperor Wen sent a scribe to visit the aged scholar. As Fu dictated the *Book of Documents* from memory, the scribe transcribed the old text in a modern idiom. When Fu's dialect was impenetrable, local people were asked to interpret it. Such were the accusing stories that emerged after the fall of the Qin.

In fact, the dynasty *had* preserved many of the classics, usually for the official palace professors. To handle the enormity of its empire, the Qin rationalized, unified and nationalized the scripts of China into an armoury of characters that could be deployed throughout the land. Written texts proliferated. Moreover, the state was divided into thirty-one counties and in each a magistrate would send written accounts back to the capital. A centralized bureaucracy had been born, stretching hundreds of miles in all compass directions, and ruling over a population of 40 million. Its currency was writing.

Although the Qin encouraged bamboo scholarship of a kind, it was only under very tight control. Nevertheless, literacy appears to have trickled down the social scale with soldiers able to read (and perhaps write) lists. Two letters dating to the third century BC were written on wooden boards by a Qin soldier fighting in a southern campaign; he asks for cash and for his mother to bring some clothes to the front line for him, and he asks after his mother's health and his wife's loyal service to her father-in-law. These letters suggest literacy at quite low social levels, as well as some kind of postal system that was available to soldiers, if not also to civilians. There is even evidence for some literacy among women – certainly among the elite, generally for women who were heads of households and possibly further down the

5 Text as cultural patriarch: the blind scholar Fu Sheng dictates the *Book of Documents* from memory to Fu Nu, his scholar-daughter, who then translates his dialect into mandarin for the kneeling scribe. (Nothing here is more fanciful than the scribe's use of paper, centuries before it became the surface of the classics.) The composition probably dates to the late fifteenth century, some 1,700 years after the dictation was believed to have taken place. Du Jin, its Ming dynasty painter, here employs a trademark formalism well suited to a scene that speaks so eloquently of one of the highest priorities of classical Chinese civilization: retention of the classics. (© 2014, Image copyright The Metropolitan Museum of Art/Art Resource/ Scala, Florence)

social scale too. *The History of the Han Dynasty*, which is among imperial China's best-loved official histories, was written by Ban Gu and his sister, Ban Zhao.[11]

Yet the dynasty would not last. After the death of its first emperor, the Qin quickly fell, the victim of popular revolt. The entire dynasty endured for just fifteen years. Importantly for the future of books and paper, the ensuing Han dynasty, founded by a rebel peasant, absorbed the horror stories of Qin misrule, preserving Qin history as a warning of the consequences of bad governance while Han scholars sifted through that history at their desks. The new dynasty took on the Qin bureaucracy, the Qin empire and many of the Qin laws, but appointed itself, the Han, as true author of the empire. As it did so, an armful of old books became the guiding stars of Chinese education, governance and philosophy. The Han founder, Liu Bang, displayed little initial enthusiasm for the empire's resurgent bookishness. An early conversation with one of his advisers, Lu Jia, and recorded in the *History of the Han*, made clear Liu's feelings.

> Liu Bang: I took the empire on horseback. Why should I bother with the *Book of Odes* and *Book of Documents*?
> Lu Jia: You took it on horseback, but you cannot rule it from horseback.

In fact, Liu would prove to be an excellent ruler and a good student: a meritocrat, a peasant in spirit and a Confucian in practice. He protected some of his weaker subjects, such as farmers, from exploitation and cut the volley of Qin laws down to three clear basic principles. He also introduced the beginnings of an empire-wide civil service recruitment system, one which chose state bureaucrats based on merit, not background or connections.

Liu's successors moulded his Confucianism into a new form of statecraft, gathering the classics together until books (most of them on bamboo, some on wood, a select few on silk) were said to be 'piled up like hills'. An Imperial Library was founded and summaries of all its books were commissioned. Classification systems were soon needed to create dozens of sub-divisions: 16 works by the Yin-Yang school alone (a school which emphasized the supreme power of elemental forces of nature) were held by the library, 249 chapters or facsimiles in all. The library held 70 different works on the 6 classics – 2,690 chapters. Two-thirds of the 667 titles listed on the Han library catalogue have since disappeared, but several Han titles dug up in recent years are not even mentioned on the list. Bibliographies and new classifications appeared too: the six classics, the various philosophers, poetry and rhyming prose, military studies, science and occultism, and medicine. One such bibliography listed 600 titles comprising more than 13,000 volumes. (Once fully edited, texts would even be transferred onto silk – an exceptionally expensive product – for the Imperial Library.) Beyond the palace and government, wealthy men bought and collected books for their own libraries.

As reading spread, talented but socially inferior civil service candidates began to work their way up towards the court. (Officialdom expanded too, reaching 130,285 members by 5 BC, most of whom would have been functionally literate. Perhaps twice as many officials again were employed at local level.) Thus, in 140 BC, when Emperor Wu personally examined more than a hundred recommended young scholars, many of them were from poor families. Before a candidate appeared, the emperor was handed a few bamboo slips with policy questions brushed onto them. One of the poorer candidates to appear that year was Dong Zhongshu, who came from the riverine flatlands south of what is now Beijing. Dong's answers, coupled with his essay arguing for a Confucian state, won him a senior role.

It was Dong's advice which led to the founding, in 136 BC, of a Commission for the Recovery of Classical Texts and, twelve years later, a Grand Academy. The latter began with 100 students but, by AD 140 that number had grown to 30,000. Throughout this period, colloquy after colloquy was held to decide which of the classics deserved a dedicated 'chair' (an official position which gave special prominence to the associated text). Writing, studying and collecting books became a state preoccupation; even in Dong's day the *Spring and Autumn Annals* had five main schools, each armed with a commentary of its own. But this culture of the bamboo text was also a logistical headache. In AD 25, porters needed more than 2,000 carts to transport the imperial book collections, and thousands of bamboo works were lost on the journey.

Dong himself believed texts were a spiritual matter. He argued the future was divined not with turtle shells, bones or yarrow plants, but with the right reading of the past. The Qin disgusted him. Asked whether Qin governance just needed renovating, Dong replied that you could not carve a piece of rotten wood or trowel a wall of dried dung. Instead, Dong wanted the classics cut to five texts and, in the mid-130s BC, he got his wish. From this point until AD 1, wrote one of the Han historians, it was not uncommon for a scholarly study of a single classic to run to a million words.

Some, however, viewed this outpouring of written opinion as prolix. In the first century AD, the great philosopher Wang Chong began life in the Chinese capital in such poverty that he had to memorize the classics by reading them at the bookstalls. Later in life, he complained that scholars merely pursued their studies in 'old ruts'. A commentary on the first sentence of the *Book of Documents*, one of the Confucian classics, was 20,000 characters long. The *Book of Changes* and *Book of Documents* both attracted commentaries of 300,000 characters each. (By the eighteenth century, commentaries on the classics were a hundred times longer than the classics themselves and stretched to 50,000 scrolls.) Yet even in first-century AD China, this over-writing had its high-profile critics, among them the renowned official historian Ban Gu, who lamented the nit-picking extremes to which writers had gone in their commentaries:

> But they only deceived themselves. That is the tragedy of scholarship.

Ten miles north of Xi'an a hewn mound rises a hundred feet above some of the world's oldest farmland. From its flat summit, which spans some 260 feet, you look down over vineyards, apple orchards and pine farms. Elderly men fly kites on the crest, surrounded by treetops and mist. Below the mound, a staggered series of platforms leads south to a driveway flanked by trees. There are two sheep tethered to one of the trees. This sheer and obscure mound is the last vestige of Weiyang Palace, where scholars gathered to debate the classics and so forge a canon for China's future. They began their debates in the Stone Canal Pavilion and the arguments lasted two years. More debates followed through the 50s. But it was in AD 79 that the scholars convened for their final joust, to dispute the texts at the heart of their traditions.

They discussed sacrifices, the rituals and institutions of marriage, ancient legends, emperors and gods. They explored questions like cultural refinement, mystical symbolism, executions, the eight winds (acupuncture points) and the three rectifications (which related to setting the calendar). They considered instinct and emotion, rites and music, the three major relationships (lord–subject, husband–wife, father–son) and six minor relationships, trade and the gifting of land to a feudal lord. The scholars conjured a universe of Chinese life in the hope they could funnel it wholesale into their interpretations of a small set of classical texts. Policymaking had assumed the form of debates over the texts of Confucius. In the second and early third centuries, the later years of the Han dynasty, some authors even began to argue for their right to oppose the emperor's court, especially when that court was perceived to have come under the power of self-seeking court women and eunuchs.

Zhou thinkers, Qin unifiers and Han scholars had transformed China into an empire centred on a small handful of difficult classical texts. Reading, scholarship and writing had blossomed and the civil service, the natural home of imperial China's scholars, had increased in number into the tens of thousands. It was an unprecedented transformation, not simply in China but in human history. There had never been such demand for the raw materials that at that time constituted books: ink, brush and bamboo.

Bamboo had turned plants into the surface of knowledge, a dominance not challenged until the digital age. What matters for characters, as for letters, is not simply their inky forms but also the inkless spaces between the strokes, and for at least a millennium and a half, those spaces were predominantly bamboo. Bamboo oversaw an outpouring of history writing as well as of bureaucratic records. From the late eighth century BC until the downfall of imperial China in 1911, official records were made for almost every year of China's history. This was a legacy of both the Zhou dynasty and of bamboo itself.

Bamboo oversaw the birth of a new imperial power in East Asia. The strength of this new empire lay not simply in munitions and money, but also in books and the literary culture that would act as both a buttress and counterpoint to political authority. Bamboo was celebrated by China's poets, one of them calling it 'our elegant and accomplished prince ... grave ... and dignified' in verse written 3,000 years ago. But the clearest evidence of its transformative role lies in the graphic history of a handful of Chinese characters.

Bamboo grows rapidly, especially in China's subtropical southern regions, and the Zhou harvested it eagerly. Used during the time of the Han for houses, suspension bridges, carriages, and bows and arrows, it needed close treatment to be turned into a surface for writing. It was cut down into strips and fire-dried to evaporate the 'fresh juice' which otherwise brought decay and attracted insects. Most strips were less than an inch wide and little more than a foot long, allowing a scribe, sitting cross-legged, to hold one between his stomach and his left hand. This favoured writing in columns not rows, something not changed in China until the 1950s. (Even the standardization of starting on the right, not the left, may derive from how a scribe laid out his bamboo strips to his left, one by one, as he finished with them.) Binding a few strips together with cord to form a longer text forged the character *ce*, which still means 'copy' or 'volume':

Mistakes made on bamboo could have made writing exceptionally time-consuming, since a whole prepared strip might need to be discarded. To remedy this, book-knives were developed, and some were elaborate enough to be 'etched with the name of the manufacturer', according to Lu You, a poet living in the second century AD. The two-character word for 'delete', *shanchu*, reflects these origins. The first character, *shan*, includes a radical, the part of the character which indicates its approximate meaning, on the right. That radical, which appears simply as two vertical lines, one tall and one short, means 'knife'. The left part of the character *shan* refers to bamboo writing strips:

Weight, format and size, however, would prove to be bamboo's weaknesses. Bamboo books were ill-suited to referencing and to transportation, which made them more appropriate for the libraries of the elite than to a wider market. Set out on bamboo slips laid out side by side, the Bible might stretch to over 600 feet and would probably need a cart to transport it. It was in the bamboo age that a learned man would be described as having 'five cartloads of learning'. Bamboo learning could not be taken lightly.

Moreover, although a knife could be used to scrape away the occasional error, this did not mean multiple drafts were feasible. Instead, those charged with any kind of official writing, especially for the court, might have to plan their composition out in their head in advance, so as to be sure it was of the correct length and had no errors in it. Thus, the famous poet Yang Xiong (53 BC–AD 18) described the pain of having to compose a poem at the emperor's request:

> Thinking hard about it, I fell asleep as soon as I completed it. In my dream, all my internal organs had been pulled out and I stuffed them back inside with my hands. After I woke up, I had asthma for a year. The experience taught me that excessive thinking was bad for my soul.

6 Yang Xin's 樂 (pronounced 'le'), meaning 'happy' or 'happiness'. Where the white of the page dominates in 'Spring', here the ink all but obscures the centre of the page, the barely restrained pressure of the brush strokes expressing a surge of joy.

49

This cannot all be blamed on bamboo, of course, and many works on bamboo ran to tens of thousands of characters. Yet in a case like Yang's, when the request came from the emperor himself and the end-product therefore needed to be immaculate, bamboo had obvious drawbacks.

In AD 183 there was a traffic jam in Luoyang, the Chinese capital, as thousands of carts jostled to reach the Grand Academy. The forty-odd stone tablets of the Chinese covenant stood on the lawn outside, set out in U-shaped rows, the end of a centuries-long argument over which texts should constitute China's canon. The agreed text of the Chinese classics was etched out in more than 20,000 characters on these stone faces; this was a landmark moment in the history of China. Until then, only graves and war memorials had merited stone. The gathering crowds, arriving on foot or by cart or horse, did not just look at the tablets or try to remember their contents. They began to copy them too.

Less than a decade later, the capital was burnt and ransacked, and its libraries were destroyed. Cart after cart of bamboo books from the imperial libraries were taken further west but the official histories recorded that, on the way, bamboo writings were lost and silk writings were converted into covers or bags. Books still travelled badly.

Yet the idea that books were worth transporting at all was due to the elevation of their status under the Zhou, Qin and Han dynasties. The rise of the Zhou had embedded writing in government, and the dynasty's ensuing disintegration had raised the question of how to govern; scholar after scholar had replied with his suggestions, written out on bamboo or wood. As a result, the Zhou had given China its sages. The Qin dynasty had spurred civil service writing through its determination to standardize the empire into a single entity, producing an enormous bamboo bureaucracy and a unified script. Finally, the Han had supplied China with its editors, librarians and commentators, growing the national vocabulary to almost 10,000 characters. It had also cemented the place of Confucius; henceforth all civil servants would study his writings.

Bronze and, most of all, bamboo had been crucial partners for writing. But the growth they enabled also moved writing beyond their control, and towards the greatest of all its patrons.

4

Genesis

Had I been born Chinese, I would have been a calligrapher, not a painter.
Pablo Picasso, in conversation with the calligrapher Zhang Ding in 1956

In the middle of the fourth century, a century and a half after the fall of the Han, the master calligrapher Wang Xizhi sat in his study painting characters onto a blank page. As a child, his speech was hesitant and unclear, but when he grew older Wang found he had a genius for invention and scholarship; when he brushed out the same character a second time in a piece of writing, instead of merely rote-copying it as was expected, he would always develop its form, giving it his own particular slant.[12] In adulthood he had worked as a civil servant; at court, he was offered the secretariat of the Ministry of Rites. But he preferred to resign instead, citing ill health. He retired to the city of Shaoxing to wind down his career in the riverine flatlands of northern Zhejiang, a little south of the Yangzi River.

At home he reared geese and fathered seven sons; each of them would turn calligrapher, too. People said that it was watching a goose bend its neck which taught Wang, who would become the author of China's most legendary piece of calligraphy, how best to turn his wrist as he held a brush to the page. He painted characters not simply to pass on knowledge and records, as the Han had done, but for their own sake. His running script, in which each character is drawn with the brush only rarely leaving the page, formed an electric wire down the columns of text, dynamic and humorous. He shunned tradition on the page as he did in public life.

Wang, known even today in China as the Calligraphy Sage, treated his passion as an end in itself. The art became known as 'music without sound, in which the brush dances and the ink sings'. Writing was not simply a means for spoken words to be passed from one mind to many: that would reduce writing to a mere convenience. Instead, the craft of writing was valued in its own right. Brushes allow a scribe far greater freedom of expression than pen nibs, and the rise of the 'running' and 'grass' scripts in the Later Han dynasty, both of which involved the brush leaving the page much less, served to increase the freedom of the scribe more than ever. With that greater power to fashion a text, the status of the scribe and his art would rise. In the tenth century, a group of Chinese scholar-officials would go further still. They argued that the purpose of painting itself (an art less distinct from calligraphy than in the West because both writing and painting were done with a brush) was not to recreate what already existed, but to enable the artist to express himself. This emphasis on self-expression in China's visual arts had begun with calligraphy.

Wang's desk would surely have been covered in the writing paraphernalia of the day: a carved ivory or bamboo arm-rest, a stone seal-stamp, a jade brush-wash, a birch wood or porcelain brush-holder, a jade paperweight, a sculpted brush-rest and a Sichuanese book-knife. But there were four items he could never do without. Those four were a bundle of metaphors for civilization's oldest game. As he explained:

> Paper stands for the troops drawn up for battle; the writing brush for sword and shield; ink stands for the soldier's armour; the ink stone for a city's wall and moat; while the writer's ability is the commander-in-chief.[13]

Wang valued the 'four treasures' in their own right, like every Chinese scholar. A surviving set of them, kept in a bamboo box, dates from 2,000 years ago. By the second century AD, brush and ink were given to high officials every month. A century later, the four treasures were gathered for use at the ceremony used to appoint a crown prince. A writing set was the indispensable toolkit of the Chinese gentleman, civil servant and ruler. Only speech itself was reckoned to be as important for a civil servant:

a Han scholar wrote that wise and able men would be most useful to the court by using their three-inch tongues and their one-foot brushes.

Wang's ink was made with the soot of burnt pinewood. In a fifth-century recipe, blue pinewood was pounded and the mix strained through a sieve of thin silk. Wang might have then added glue, perhaps made from fish skins, in order to dissolve the bark mix. In the first millennium AD, pine, soot and glue formed the base mixture; thereafter lampblack ink (made with animal, vegetable or mineral oils) became more popular. For the best ink, five egg whites, an ounce of cinnabar and the same quantity of musk were each strained and then added to the solution. Other precious ingredients included pearl powder, the skins of pomegranates and pig's gall, which improved the lustre of the ink's finish on the page. The mixture was beaten down in an iron mortar until it formed a dry paste, but it had to be pounded at least 30,000 times before it formed a cake ready for use. Wang would have used small ink sticks of two or three ounces, hard as jade, effulgent like lacquer, possibly scented and always purest black.

Ink was especially revered. The British sinologist Joseph Needham (who probably used more ink than any other Westerner in writing on China) commented in his multi-volume *Science and Civilisation in China* that the names of hundreds of Chinese ink-makers are recorded in Chinese literature, whereas very few of the empire's papermakers or printers are known to us by name. In the West, the substance became known as Indian Ink, and the Roman historian Pliny the Elder wrote in its praise. However, it is more likely to be a Chinese invention, given the number of pre-Christian era writings in 'Indian Ink' found in China, the equivocal (or outright scornful) attitude to writing expressed by some of the more ancient Sanskrit texts, and the fact that Chinese goods often travelled to Europe through India in the last centuries BC and first centuries AD.

Ink's life was measured in months, but the life of the third treasure of the study, the ink stone, was counted in generations. (Scholars even coined evocative names for individual ink stones, like 'monkey reaches for peaches' and 'female phoenix gazes at the sun'.[14]) Clear water sat in a cavity on its surface, beside a flat area where Wang would rub the ink stick, gradually creating a sticky residue. When a viscid pool had appeared on the ink stone, Wang would pick up his brush.

Cutting a bamboo to make a holder,
Tied with silk string, covered with lacquer.[15]

Such was the shaft of the brush, used in China on pottery as early as the fourth millennium BC. Brush hairs were first made with the feathers of chickens, geese or pheasants, the hair of rams, hogs, tigers, perhaps a man's beard or even, some early references claim, the hair of a newborn baby. By the third century BC rabbit hair formed the bristles, with a tip of long and tough hairs that encased an inner tuft of shorter hairs. Rubbing an ink stick onto an ink stone and dipping a brush into the ink that it formed was nothing new. Wang might have been doing the same 500 years earlier. But since then, authors and calligraphers had multiplied and, in the process, they had added a new

7 Maceration: steaming the bamboo fibres in central China as part of the papermaking process. Beneath the cistern a fire is fed to heat a water boiler. This 'steam bath' effect (as the photographer described it in his diary) loosens the bamboo, allowing fibrous cells to be detached. The photo was taken by the Roman Catholic missionary Father Leone Nani, who left an extraordinary photographic record of his time in rural China (1904–14). It was during Nani's years in China that imperial rule finally ended, a system which had begun close to the time paper first emerged, more than 2,000 years earlier. (Copyright © Archivio PIME)

technology to the tools of their trade. Thus, when Wang picked up his inky-bristled brush, it was to set his calligraphy down on the fourth treasure of the study: paper.

The paper trail begins with a few isolated scraps unearthed across China. A thick silk had been used for writing in the final centuries BC and it gave rise to another silk, made from leftover cocoon hairs. The new writing surface was called *xiti* ('she-tee'). *Xiti* was made from animal fibres, a cousin of paper but not paper itself, since it is not strictly made from plant or vegetable fibres and is not macerated. That plant or vegetable derivation is paper's first defining characteristic. However, the Chinese character for paper comes not from the nature of paper itself but from silk. On the left of the character is the 'radical' that means silk; a radical is a pointer to the meaning and, in this case, it looks like a capital *E* underlined. The right part of the character is simply phonetic. The Chinese word is *zhi* ('jerr') and the tone falls and then rises like an upturned Chinese hat.

纸

The confusion with silk muddies the early history of paper but *zhi* seems only ever to have been used to mean 'paper'.[16] Both terms were used in the official dynastic histories, as on the note written on *xiti* and sent in 12 BC from Emperor Cheng's wife, the singer and dancer Zhao Feiyan ('flying swallow'), to the maid-in-waiting who had recently borne the emperor a baby boy. (The boy had since been killed and the note requested the maid's suicide.) Or, still earlier, in 93 BC, when the dying Emperor Wu's spymaster Jiang Chong advised the crown prince to cover his nose with a piece of *zhi* when visiting the emperor, because 'the emperor hates big noses'. In a dictionary produced close to AD 100 by the scholar Xu Shen, *zhi* is defined as 'a mat of refuse fibres', a clear description of paper, not silk. The fourth treasure of the study had entered the official lexicon.

*

In 1975 a tomb was excavated in Shuihudi in central China. It dated to 217 BC and one of the bamboo documents found inside was a day book, in which the word 'paper' makes its first appearance in history.

> If a man's hair without reason stands erect like worms, whiskers or eyebrows, he will have encountered a bad spirit. To resolve this, boil a hemp shoe with paper, and the evil will be dismissed.[17]

In the same century, a model horse was sculpted and placed in a tomb in southern China. On its back was what has been pronounced the oldest known scrap of paper in the world, though its age has never been independently tested. More scraps have been unearthed from the second and first centuries BC, mostly from China's north-west, where arid conditions have made a museum of ancient rubbish dumps. Some of these scraps were used as wrapping. On their crumpled surfaces are the names of the medicines they encased.

It is no coincidence that paper makes its first, fragmentary appearances in the poor and distant north-west of China rather than close to the wealthy cities of the Yellow and Yangzi river basins. Bamboo and silk were the writing surfaces of the rich; they were dignified, worthy of the brushes of aristocrats, senior civil servants, educated poets and emperors, and these people all lived in the east and centre of the country. Paper, on the other hand, was not yet associated with learning or status. Moreover, while bamboo grew widely across most of China, it did not grow in the far north or north-west. Yet, it was the dry climate of the north-west that was best suited to preserving paper, rather than the humidity of the south.

Paper may have been a haphazard discovery. Accounts survive, from the third century AD for example, of old women washing coarse silk before bleaching it. A washerwoman might spend twenty or thirty days bleaching her silks after washing them. Treating refuse fibres in this way was already common in the third and fourth centuries BC. Had they been left to dry on a flat surface for long enough after being washed and bleached, they would have made a good surface for writing. Many Chinese scholars and nobles wrote and painted onto fabrics already, so this need not have seemed a significant shift at the time.

For want of any evidence, however, there is only speculation. Perhaps a tradesman, looking for new curios to sell, invented paper to use as decorations, clothing, furnishings, windows, kites, cleaning aids or to make models or toys. In an empire of engineers and writers, paper might have been invented several times over. No historian today is confident that it was first invented as a surface for writing. But paper's first use as a writing material matters far more than the moment of its invention.

Historians have at least provided us with a figurehead to look back to. Cai Lun did not invent the concept of setting brush to paper, but it may well be that Cai, and the empress he served, were the first to appreciate its potential. They are rightly placed at the beginning of paper's rise.

Cai Lun grew up in the late first century AD in Leiyang, a town in Hunan, the southern province which skirts the south bank of the Yangzi. The area had become an immigration hub for ethnic Han Chinese from the fourth century BC, as droves of migrants moved there to plant rice, clearing many of the forests as they arrived.

Cai's career as a civil servant gained momentum in AD 75, when he became a liaison officer with the Privy Council in the capital city of Luoyang, a job that included acting as chamberlain to members of the royal family. Within two years he had become political adviser to the emperor, chief of the Imperial Supply Department (which stocked all palace stationery), and imperial craftsman, entrusted with making furniture, swords and other implements for the imperial household. It was during this period that he began to study how to produce paper efficiently.

In 89 or 90, the Emperor He made an inspection tour of the Imperial Library in Luoyang. There he met Cai, a eunuch[18] who had entered the apartments of the imperial harem late in the reign of the former emperor. Soon after Cai's arrival, an obsessive scholar was summoned to court. The new scholar had an unusual passion for books, and recommended that vassal states sent paper to the emperor as tribute. It is hard to imagine he and Cai did not discuss the technologies of writing together. By 89, Cai was director of the Palace Workshop, where new designs for instruments and weapons were proposed. When the emperor saw the hugger-mugger of the Imperial Library, its wooden boards stacked up in piles and its contents in disarray, he asked Cai Lun to study and order them. But Cai needed a better material with which to make records.

In Leiyang, the warm southern town where Cai was raised, the humidity favoured papermaking more than in the arid north. Ramie, a perennial nettle indigenous to East Asia, is especially common in the north of the province, and waste from its processing in cotton factories is still used to make paper. It was used for fabrics before Cai was born, and the process of felting plants like ramie could even be what suggested the possibility of papermaking to him in the first place. (Felting is the process of mashing, pressing and condensing fibres.) In Cai's time, local people used the bark of the paper mulberry tree to make clothing, covers and wrapping. By adding water, the bark could be pounded to ten times its original size, thus creating a surface quite similar to paper, and pieces could be glued together, end on end. Working alone, a manufacturer could fabricate only two or three sheets a day – it was a laborious undertaking.

Cai chose felting over pounding. He engineered a plant-fabric paper made from mulberry bark, hemp and rags, together with flax, cloth and fishing nets. Although he began with the paper mulberry tree of his southern home, the barks, plant basts, grasses and other fibres that would henceforth be used to make paper were legion: mulberry, bamboo, ramie, hemp, straw, cabbage stalks, thistles, St John's wort, turf, mallow, lime tree bark, corn husks, genista, pine cones, potatoes, reeds, horse chestnut leaves, walnut tree bark and jute, to name a few.

Cai mixed the bark, hemp, rags and flax with water, possibly after bleaching them, then mashed his ingredients to a pulp, pressed out the liquid and hung the resulting mat up to dry in the sun. A scrap of woven cloth stretched across a bamboo frame would have acted as effective gauze, leaving a sheet of paper on its surface, ready to dry; this quickly became the popular means of maceration (the process of soaking the fibres in water to form a pulp). These, then, are the two elements that are needed to make paper: its raw material must be plant or vegetable fibres and it must be macerated. Other processes can look deceptively similar, most significantly 'papermaking' in Central America, which began no later than the fifth century AD, and possibly centuries earlier. Yet, although the raw materials were soaked in preparation, they were soaked so as to be pounded and matted into sheets, and not so as to be pulped. Only the latter is true maceration. Nor is this merely some excessively technical line in the sand – like its later Polynesian counterpart, the Mesoamerican product (*amate*) was not convenient for rapid, detailed and extensive writing due to its rougher surface. After the Conquistadors arrived in the New World, it was Francisco Hernández de Toledo (1514–87), court physician to the king of Spain, who noted (correctly) that *amate* was inferior to 'smooth' European paper. Cai Lun's true paper, on the other hand, was not only easy to use but quick to produce. The same worker who could deliver two or three sheets a day by pounding bark could deliver 2,000 sheets of paper by Cai's method. But if Cai was the technician who enabled paper to become a viable writing surface for mass use, it was Empress Deng who had the vision to act as matchmaker to paper and writing.

Deng Sui was the granddaughter of a Han prime minister. (Sometimes translated as 'chancellor', this was the highest administrative post and involved setting the government budget.) She was born in 81 in Nanyang in the cattle-country of the near north. By the age of six, she knew Confucius's *Book of Documents* and at twelve she had read the *Classic of Poetry* and the *Analects*, according to her official biography. While Deng's mother complained that she did no housework, the young girl used her nights for study instead, and became known as 'the scholar'. Deng's father realized she was far more able than any of her brothers, and he often discussed business with her. She even received spiritual sanction when a physiognomist told Deng that her face had the same contours as the fabulous founder of the Shang dynasty, King Tang.

In 95, she joined the imperial harem as one of its best-looking women. Enthusiastic Han historians wrote that Deng was seven feet two, that she was courteous and modest and that she dressed plainly for banquets. She fascinated Emperor He too. When the empress died in 102, Deng was chosen to take over as leading lady of the empire.

In 106, while China was in the midst of a financial crisis, the emperor died and Deng took his place; she ruled the empire for the next decade-and-a-half with conspicuous competence. (Female supreme rulers were a rarity in imperial China.) Twice she opened the imperial granaries to feed the hungry; she repaired waterways and cut court ritual and banquets. She ate only one meal a day, reduced the fodder for the imperial horses, slashed marquisate revenues (the income landlords received from the land they rented out), cut manufacturing and sold offices and titles. She dealt with serious rebellions in the west and the south, as well as crippling floods, droughts and hail storms in several parts of the empire. She possessed an exceptional aptitude for the stratagems of territorial politics too, but her legacy was, above all, in the arts. She remained bookish throughout and opposed the tired formulas of official teaching. She created new positions for scholars with high values: 'Sincere and Upright', 'Plain and Honest', 'Kind and Worthy'. Even as empress, she maintained her interest in classics, history, maths and astronomy. She even called seventy members of the Deng and imperial families to study the classics together, and oversaw their examinations herself. One of the major projects sponsored during her rule was the standardization of the five classics, and she chose Cai Lun to supervise proceedings in the Imperial Library, situated in the Eastern Pavilion of the Imperial Palace.

In 114, the empress dispensed with the department Cai ran and appointed him a district governor. He became her coachman and probably ran her household too. According to the voluminous *Comprehensive Mirror to Aid in Government*, an eleventh-century polymath's three-million-character account of fourteen centuries of Chinese history, Deng refused all tributes from abroad (from today's south-west China, Vietnam and Korea, for example). Instead, she insisted on receiving annual gifts of paper and ink. The official *History of the Han* records a similar tale. (Evidence for such paper tributes arriving from south-east Asia survives from the third century onwards.[19])

She encouraged Cai in his paper experiments, monitoring his progress and funding his research. Following his success, Deng promoted 'the paper of Master Cai' at court, which marked paper's first appearance at the highest levels of the empire; it had previously been viewed as beneath the elite. (Indeed, a letter survives from decades after Deng's death in which an official apologizes for writing on paper.)

Cai's paper evolved through careful analysis and a series of experiments; his predecessors had not created anything so adept for brush and ink. Building on what Cai and Empress Deng had begun, China's imperial government library department at Chang'an eventually set up its own paper mill in the ninth century. The mill sat under the authority of a committee (some historians have likened it to a guild) of four expert papermakers.

Deng died in 121, the same year that Cai's paper received official sanction (and with it, more widespread use), although the government continued to use bamboo as well for some decades. Following her death, Deng's enemies at court hounded her family and household with accusations of treason, and they ordered Cai to report to the Minister of Justice. He bathed, put on his robes of state, and drank poison.

Cai's legacy was a new kind of paper, made from plants that grew rapidly and spanned much of China. Through the rest of the second century AD, paper gained in quality and dropped in price, becoming widespread for the first time thanks to Cai's craftsmanship and Deng's foresight.

Paper still had to compete against the status conferred on bamboo and its associations with the most venerated writings, from the classics to government histories to poetry. The fact that bamboo's very shape (when cut into long vertical strips for writing) fitted neatly with literary conventions only added to the sense that paper was less appropriate for serious writing. Silk was, likewise, a more respectable surface, in part because of its association with quality and so with the lives of the elite. Twenty years after Cai's invention, the scholar Cui Yuan wrote in a letter (doubtless delivered by the government postal service, in place since the Zhou dynasty) to his friend Ge Gong: 'I am sending you ten volumes...They were written out on paper because I am poor and unable to do them on silk.'

After the Han capital was moved back to Chang'an in AD 190, the final three decades of the Han dynasty were disordered and unpredictable. With political and social disintegration came paper's great opportunity: they once again led the elite to question the very foundations of the dynasty, including its Confucian orthodoxy. Amid the mayhem, it was hard to preserve the old bamboo and silk system, a system so tied to the traditional uses of writing (as to its sometimes terse and ambiguous writing style) and to the political order of the Han. Bamboo was naturally identified with the values presented for centuries by the writings on its surface. In its dimensions, it could also set limits on how much an author said. Moreover, in a time of social upheaval, a bamboo library became a vulnerable asset. It did not travel well.

Paper, on the other hand, was easier to prepare and better able to accommodate wordiness than bamboo; its versatility extended to its relative freedom from limitations on writing styles. Paper was a blank page. It allowed an author to write uninhibitedly and express his feelings and thoughts as fully as he wished. He could revise his drafts more often than before. He could even fit a long work on one single sheet of paper by folding it up into several sections. In this way it made it easier for him to maintain a narrative plumb-line, to hold onto the thread of his argument or focus throughout – not that he had not managed this before, of course, but paper presented everything more immediately in front of him. An author no longer needed to plan characters all out in advance in his head. He had more space to express his feelings in detail and to explore new forms of writing. Paper gave him greater control over his written work.

Paper allowed friends to converse with one another despite distance. Through the decades of bamboo and paper's co-existence, intellectuals continued to write formal government reports on bamboo, but they also wrote letters to one another, initially on wood or bamboo, but from the early second century AD on paper. (The shift may have been smoothed by letters' mediocre literary status; the biblical canon features twenty epistles – the Confucian canon, not one.) Some of these second-century letters expressed the wonder many intellectuals felt at being able to communicate with friends easily (and inexpensively) even when they were many miles away. Ma Rong and Dou Zhang, living in the second century AD, were two such friends, separated by distance but kept close by paper. The official *History of the Later Han* recorded the exchange:

> Ma Rong wrote in his letter to Dou Zhang that he was very pleased to read Dou's letter since it felt like meeting face to face.

For men like Ma and Dou, paper allowed them to communicate better with other artists and scholars and to read and contribute new writings more easily. Just as paper allowed authors a more independent voice and a higher word count, so it also enabled them to listen to the voices of many others, even those hundreds of miles away. The official *History of the Jin Dynasty* tells how the great poet Lu Ji (265–316) was frustrated at not hearing from his family, and communicated his frustration to his dog, Yellow Ears, who growled and wagged its tail. Lu Ji then placed a letter in a small bamboo case, fastened it to the dog's neck, and was later pleased to find the dog return with a reply, having travelled to his family's home and back. This technique, the *History* boldly claims, then grew 'quite common'.[20]

Paper letters were soon joined by prose-poems on paper; both these forms blossomed after the fall of the Han in 220. In time paper would also begin to claim the classics, but before that happened it started to drag some less elevated forms into the literary spotlight too, allowing literature once thought low-brow to win new appreciation, including less traditional forms of verse, which could be more expressive or more playful.

While paper books grew in popularity through the second century AD, paper only began to gain the edge over bamboo in the middle of the third century, as greater production reduced its cost and improved its social standing. But if the price of Cai's paper had much further to drop, it was already noticeably cheaper than bamboo; scholars and civil servants were certainly using it as early as the second century, especially for writings of a personal nature, presumably in part because they had to fund the materials of such writings themselves. By the early third century, there are recorded examples of paper being used as stationery by civil servants. Within little more than a century of Cai and Deng's passing, therefore, China was entering its paper age.

By the fourth century, when Wang Xizhi was growing up, China had become a paper culture and it was in this paper culture that he would learn his craft. In 352 the calligrapher was fifty years old. He invited forty-one friends, among them poets and other calligraphers, to accompany him to the annual spring purification ceremony (when cleansing rituals were undertaken to prevent illness and misfortune) at the Orchid Pavilion in Shaoxing. The group sat on the banks of the stream by the pavilion to wash themselves free of bad luck. Servants released small lotus-leaf cups of wine down the stream towards them. Whenever a cup paused in front of one of the group, the guest had to drink the contents or write a poem. By the end of the day, Wang and his friends had written thirty-seven poems.

Wang was already half drunk when he decided to write a poem. He picked up a brush made from rat hairs and whiskers and brushed out *Preface to the Orchid Pavilion* on silk in the 'walking' style, a style at once smooth and busy, creating a rhythmical stream of characters. It was his best work and is still regarded as the finest piece of calligraphy in China's history. Scholarship on this one piece of work has not ceased through the centuries, though since the mid-seventh century it has relied on copies when the original was no longer available. Wang subsequently made more than a hundred copies of the *Preface*, but none matched the elegance and vivacity of the original. Guarded for centuries, an emperor finally obtained it through trickery in the seventh century and had it buried with him when he died, in 649.

The successful marriage of writing and paper was germinal to the spread of knowledge. By the fourth century, scrolls were no longer rare objects. Instead, they were affordable and, in time, they even grew commonplace. With that shift, knowledge no longer needed to be in the sole possession of the elite. With this release came the proliferation of paper on an extraordinary scale. Herrlee Creel, a linguistics historian, estimated that, from the second half of the second millennium BC, when writing was almost fully developed in China, up until the mid-eighteenth century, more books were published in Chinese than in all other world languages combined.[21] The empire's size and bookishness partly lay behind that, but it would have been impossible without the early rise of paper in China.

Wang Xizhi had many of his scrolls of calligraphy buried with him. Indeed, Wang's letters have provided his largest calligraphic legacy. His writing communicated meaning through the shape and personality of the brushed characters as much as through their dictionary definitions. Chinese calligraphy fused meaning and mood and it signalled the growing importance of the writer as someone more than a mere scribe. Paper had made regular writing and experimentation possible. Passionate, idiosyncratic and elegiac in his writings, Wang was a founding member of the new, paper age.

5

At the Margins

I do not like rolled-up manuscripts. Some of them are heavy and smeared with time, like the trumpet of the archangel.
Osip Mandelstam, 'The Egyptian Stamp'[22]

In 1907, in China's remote north-west, a Buddhist monk and a Hungarian archaeologist finally lifted the lid on the early centuries of China's paper age.

Seven years earlier the monk, Wang Yuanlu, had stumbled on a hidden doorway to a sealed cave. Inside, he found bundles of old writings and drawings. Wang took a couple of the papers he found there to show to the county magistrate, Yan Ze, but Yan failed to recognize their value. Three years later, the new county magistrate visited the cave and took away a few manuscripts, but merely advised Wang to look after his find, nothing more. Further attempts by Wang to receive funds for preservation brought no more progress.

In 1907 the archaeologist Aurel Stein arrived in Dunhuang with his translator, and befriended Wang. Stein had worked as a university professor, archaeologist and philologist in South and Central Asia for almost twenty years. When Wang finally showed Stein and his translator to the cave, Stein tried to suppress his stupefaction:

> The sight disclosed in the dim light of the priest's oil-lamp made my eyes wide open. Heaped up in layers, but without any order, there appeared a solid mass of manuscript bundles rising to ten feet from the floor and filling, as subsequent measure showed, close on five hundred cubic feet.[23]

Before him stood more than 42,000 scrolls and rolls, stacked high against the cave walls. Almost all were paper. More than 30,000 were Buddhist sutras and apocrypha, but there were Daoist and Confucian classics too, as well as several philosophers' tracts. (Apocrypha usually refers to extra-canonical works but Buddhism has an open canon. These are classed as apocrypha because they falsely claimed to be translations from Indian originals.) There were short stories, business contracts, poems, calendars and official papers. Stein had just been led into the greatest-known time vault for both Buddhism and first-millennium AD China. The cave had been sealed – we cannot be certain why – since 1056. Stein had come face to face with the stacked evidence of paper's first mass movement.

In 1862, Marc Aurel Stein was born into a Hungarian Jewish family, baptized a Lutheran and named after Marcus Aurelius in an apparent nod to the old emperor's religious tolerance.[24] His family lived in Pest (which only officially merged with Buda in 1873), a stone's throw from the Danube, where the local Jewry had developed a prosperous bourgeois quarter. His father, Nathan, fought in the country's war of independence in 1848–9 but Aurel's future had more to do with his uncle Ignac.

Ignac Hircschler was a classicist, the president of Hungary's Jewish community and Budapest's first eye-surgeon. He never took a Magyar name but he argued vigorously, and with all the optimism of an Enlightenment idealist, that Hungary's Jews must assimilate. Across the river lay the castellations and conservatism of Buda, but business, nationalism and the Hungarian Academy of Sciences belonged to Pest. Ignac's cosmopolitanism, scholarship and idealism rubbed off on his young nephew.

Aurel Stein studied Latin, Greek and Magyar, which he loved above all, before his family sent him to Dresden to learn German when he was ten years old. There he grew lonely, but read eagerly about Alexander the Great, who had carried Greek culture into Asia in the fourth century BC. In 1877 he returned to Budapest, now a single city, to study oriental languages, and later took scholarships across Europe. He studied in Tübingen, Vienna, Cambridge, Oxford and London. At the British Museum he studied coins from Persia and Central Asia and resolved to learn how India, China, Iran and the West had clashed and overlapped in their ancient histories.

After a decade as a philologist and archaeologist in Lahore, Kashmir and Calcutta (from 1888 to 1899), this curiosity led him to Central Asia. Here he was at last grateful for his year of military service in Buda in 1885, where he had studied topography with the Hungarian army, since it was during those years that he had learnt to map his own routes and finds. It was in Central Asia that Stein would make his greatest discovery, the discovery which tells the story of paper's first religious blossoming as little else can.

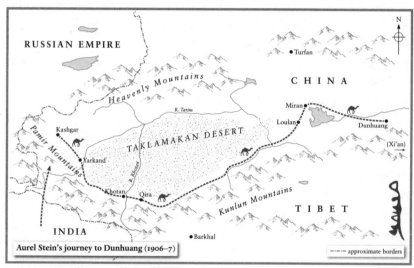

Aurel Stein's journey to Dunhuang (1906–7)

8 Few significant Eurasian religions are strange to Xinjiang ('new province'), as Chinese Turkestan is now called, a historic nexus for the exchange of goods, ideas and scriptures. Dominated by deserts and mountains, the region has often been treated as an arena for competition between great powers. Spies and archaeologists from the West (sometimes the same people) jostled for access, information, contacts and influence here throughout the nineteenth century. Today, only two major powers are widely apparent: China and Islam.

Stein was not alone in his quest to dig up Central Asia and it was news of the expedition of a German archaeologist, Albert von le Coq, which would nudge him towards his extraordinary quarry. In 1903 von le Coq was planning to visit Dunhuang, an oasis town in western China. He was from a wealthy German family and had ended up taking his expedition on behalf of the Ethnology Museum in Berlin. His interest in Dunhuang hinged on a tantalizing rumour that, hidden somewhere in the town, was a prize hoard of ancient manuscripts. Instead, von le Coq was sent to Kashgar, in the saddle of two mountain ranges and a thousand miles further west. Stein learnt of von le Coq's change of plans and decided to travel to Dunhuang, doubtless hoping to uncover whatever treasures lay there before anyone else could. But he knew nothing of the rumour.

In April 1906 Stein arrived in Yarkand, hired a local translator and a surveyor, and bought eight camels and twelve ponies in Kashgar and a horse he called Badakhshi. He carried a passport that had been stamped 'great official' by the government of India. It was not his first journey in the region but most of Stein's more severe hardships lay ahead. (In 1908 he was caught in bad weather at 20,000 feet while tracing the source of the Khotan River, and his feet were seriously frostbitten. Stein had to descend to the nearest hospital at Leh, in India's far north, where all the toes on his right foot were amputated.)

By October 1906, five animals were dead and Stein had heard news that a third archaeologist, a Frenchman called Paul Pelliot, was already at Kashgar and aiming for Dunhuang too. A race had begun, although as yet no one knew what prize lay there. Aurel began his desert crossing but met no French or Germans anywhere. In the ruins of ancient Loulan he wrote a letter to Percy Stafford Allen, the president of Corpus Christi, Oxford and a friend with whom Stein corresponded for three decades, about his departure across the desert. He told Allen that he had begun his crossing at Khotan, 270 miles along the road east of Kashgar. 'The 1,000 miles race from Khotan (done exactly in three months including one spent in excavations) is won – for the present,' he wrote.

In the ruins of Loulan, the site of a second-century BC Chinese administrative headquarters, Stein found some writing on a strip of white silk. It was written in Kharoshthi, a script used at the northern margin of South Asia from the third century BC to the third century AD. Older than any extant writing on paper, it was thus the first proof that silk had been used for writing before paper. He dug up a scrap of paper too, the surface of which carried an unrecognizable script that looked a little like Aramaic, but which he could not immediately identify.

Stein uncovered still more rarities as he travelled to Dunhuang. Almost 500 miles east of Loulan, at a site in the oasis of Miran (abandoned in the third century AD) Stein found a stucco head, 17 inches wide, and a mural of classical Greek cherubs. It was, by hundreds of miles, the most eastern find of any Gandharan art, a Greco-Buddhist style of art practised during the last centuries BC and first six centuries AD. He found evidence of the extraordinary reach of Roman influence too, when he read the identity of one of the artists written on the murals. The name was Tita, from the Roman name Titus.

He travelled on from Miran with his retinue, mostly across desert, for seventeen days until he reached Dunhuang. The only sign of people on this journey was, Stein wrote, the occasional scattering of human bones.

Once at Dunhuang, Stein headed outside the city to the Valley of the Thousand Buddhas and its decorated caves. On their murals Gandharan and Chinese Buddhas sat cross-legged, dressed in ochre against green, pale blue and lavender backgrounds. Some had been painted 1,500 years earlier. It was now, in Dunhuang, that Stein heard the same rumour that had reached von le Coq's ears a few years before.

When Stein arrived, Wang Yuanlu, who served as the guardian of the Dunhuang shrine, was making his periodic alms-seeking journey around the oasis. Unable to access the shrine without Wang, Stein instead retraced his steps to the desert where, on an earlier journey, he had discovered some 2,000-year-old watchtowers. As he explored a second time, Stein unearthed first-century BC writings on wood, some of them lists of sick soldiers living in the garrison nearby. He also found eight folded letters written in the same script as the paper scrap from Loulan. It was, he later discovered, the lost language of Sogdiana, a large Central Asian province of the Persian empire.

When he returned to the oasis, Stein met Wang, the diminutive soldier who had become a Daoist priest at Dunhuang after his discharge. Wang had arrived there in the 1890s. He had since spent his days cleaning, restoring and adding new murals to the Thousand Buddhas Caves, and funded his restoration projects by conducting alms-collecting tours of the oasis.

Stein learnt from their conversations together that Wang loved Xuan Zang, the immortalized Buddhist monk who had carried scriptures from India to China, some 2,500 miles, in the seventh century

AD. One night, after an animated discussion with Wang about Xuan Zang, Stein's translator appeared at Stein's tent carrying a few bundles of scrolls Wang had released. According to Wang, some of the scrolls had been brought to China by Xuan Zang himself.

Even such an enormous claim could not have prepared Stein for the contents of the cave, to which Wang now took him. Three hundred and eighty of the rolls were dated, ranging from AD 406 to 995, but there were other, undated papers that proved to be older still, from the third and fourth centuries. Most were written in Chinese but others were in Sogdian, Sanskrit, Eastern Iranian, Uighur and Tibetan. Some of the scripts used, like Khotanese, had been previously unknown. More than twenty languages were represented in paper piles in the cave. There were careful paintings on silk and linen, banners, and more pictures on paper and on textiles.

The extraordinary range of beliefs found in the region was played out on the scrolls Stein leafed through. There were Daoist texts, Confucian texts, Manichaean texts, Buddhist texts, Nestorian texts and even a Jewish text. There were records of the loans of goods, military passes, medical records, accounts, records of administrative omissions. There were census results, letters and folk tales too.

Wang agreed to let Stein study the manuscripts on condition that he told nobody about them before leaving China. For seven nights on end, Stein's translator carried bundles of manuscripts to his tent under cover of darkness. Then, while Stein waited, Wang toured the Dunhuang oasis to ensure that no rumours of these activities had spread elsewhere. Satisfied that Stein had not informed others in the oasis of the cave and its contents, Wang returned and agreed to release twenty more bundles of manuscripts. Stein pointed Wang to a Dunhuang painting that showed the monk Xuan Zang bringing scriptures from the Indian subcontinent. He told Wang that he would now return them to India, thirteen centuries later.

In all, Wang sold Stein seven full cases of scrolls, containing more than 10,000 manuscripts, and five cases of paintings, embroideries and other items, in return for 300 Chinese silver ingots – around US$200 or £75 in the money of Stein's time. In the years immediately after Stein's find, other archaeologists followed suit.[25] In 1910 Paul Pelliot, the French historian and archaeologist, bought 6,000 manuscripts for

500 *taels* (or $340). The cave scrolls of Dunhuang now lie in museums in London, Delhi, Paris, St Petersburg, Budapest, Tokyo and Beijing. Ultimately Stein was to become a hate-figure in China, depicted as a plunderer and a thief. Wang certainly under-charged him for his purchases, but if Stein was a thief, he was a thief who helped to preserve China's past. Wang had unearthed the scrolls, and had even ascertained their likely age, but it was Stein who first realized their broader significance. His discovery uncorked centuries of Chinese history.

9 The French archaeologist Paul Pelliot examines the contents of cave 17 at the Mogao Caves outside Dunhuang in western China in 1908. The pile beside him was already much reduced due to Stein's own acquisitions in Dunhuang the previous year. Pelliot's linguistic ability was exceptional and he claimed to have scanned up to 1,000 documents a day while in the cave. He took with him both quality and quantity: 4,174 Tibetan items and some 3,900 Chinese documents, as well as hundreds of Sogdian, Uighur and Sanskrit treasures. He returned to Paris in 1909 and was very publicly accused of trying to pass forgeries off as originals. Only the 1912 publication of Aurel Stein's book, *Ruins of Desert Cathay*, finally vindicated him. (Copyright © Musée Guimet, Dist. RMN-Grand-Palais/Thierry Ollivier)

Stein could not physically transport all the papers he would have liked; even so, his acquisitions from Dunhuang were myriad. The first batch of his purchases arrived in London in 1909. In 1920, three-fifths of the collection was transferred to India, as Stein had initially agreed, and it is now housed in New Delhi's National Museum. Stein's expedition had been largely funded by the government of India, and partly by the British Museum; many of his acquisitions were therefore split accordingly.

Before it was divided up, the Stein collection contained more than 20,000 items in Chinese script, among them 14,000 scrolls from Dunhuang, some 4,000 wooden writing slips (these are similar to bamboo writing slips – bamboo does not grow well in China's dry north-west), more than 5,000 fragments on paper from sites other than Dunhuang, and a collection of Chinese paper money from the Yuan dynasty, the same kind of money Marco Polo claimed to have seen printed. The find at Dunhuang demonstrates the extraordinary reach of China's written culture: Dunhuang lies almost 1,600 miles from Xi'an – then called Chang'an – China's periodic capital and the city at the heart of political and cultural life.

But Stein found papers in other scripts too. After Chinese, Tibetan accounted for the largest collection of writings; Stein found more than 7,000 items with Tibetan writing, including 3,000 scrolls and assorted breviaries from Dunhuang and 1,000 paper fragments from two other ancient sites, all reflecting the flourishing of the Tibetan empire in the seventh to ninth centuries AD.

There were the written remnants of cultures further to the north and west, like the 5,000 fragments of writing in Tangut, the script of north-west China's Tangut people, who migrated to the area in the tenth century, or the 1,300 fragments with writing in Tocharian, the language of another people of north-western China. There were also scripts from further into Central Asia: 2,000 paper manuscripts with writing in Khotanese, and 150 items with writing in Sogdian. There were 7,000 items in Sanskrit or Pali as well, both ancient languages of the Indian subcontinent that dominated early Hindu and Buddhist writings.

In short, Stein had discovered the crossroads he had always been seeking. In one cache of papers in a hidden oasis cave he had unveiled the overlap and interaction of the scripts and beliefs of lands and cultures as distant as Persia, China and the Indian subcontinent, as well as the many in between. There were still more distant influences apparent too, including a book of Esther (in Hebrew) from the Tanakh, the Hebrew scriptures.

Yet cultural exchange was just one of the movements that the scrolls documented. The cave documents also captured the breadth of paper's diffusion in Chinese life through at least the last two-thirds of the first millennium AD. While paper was still unfamiliar to Christian Europe in AD 1000, in China it already ranged widely in colour and quality as well as in the writing purposes to which it was put. New forms of writing, many of them religious, had emerged, as had new audiences, and paper became the principal vehicle for them to prosper.

Buddhist sutras (usually in scroll format) dominate the papers, confirming the remarkable energy of Chinese Buddhism on the page. A sixth-century sutra now stored at the British Library in London is initially written out on brownish yellow paper, but then switches to fragile white paper (still on a single scroll) as the quality of the calligraphy deteriorates. Perhaps money began to run low, leading the scribe to use a cheaper surface. Time (or patience) may have been short too, which would explain the decline in the quality of calligraphy. Or perhaps a second calligrapher finished the job. These are the puzzles of palaeography.

The yellower paper is quite fine and probably burnished. A special yellow dye was often used for Buddhist papers, possibly because of the colour's association (in Buddhism) with suffering, a theme which has always loomed large in Buddhist thought as in its scriptures. But the dye also operated as an insect repellent, aiding preservation. (This was even more important in the humid Chinese south than in the deserts of the north-west.)

The sutra's wide-spaced chain lines (lines formed in the direction of the paper's manufacture by the supports on the gauze it is dried on) are clearly visible when you hold it up against the light; they follow a slightly wavy pattern. The lay lines – narrower, closer lines running at right angles to the chain lines (also formed by the gauze) – are visible too. The paper is rough and coarse in its finish but fine instead of thick, and impressively solid in its quality.

The fibres in ordinary, modern paper have all been chopped up finely until the paper is not the 'mat' described in Xu Shen's AD 100 dictionary but more a mesh of tiny paper flecks. Thus most ordinary paper today makes a high-pitched sound when you flick it back and forth in mid-air. But early Chinese papers, like the six-foot sutra scroll's yellower pages, were of a better quality and, when you flick one through the air, make a lower and deeper sound, thanks to the length of the fibre strands and to the paper's thickness. Today such paper is usually produced only for letters or luxury uses.

This sixth-century sutra was a fairly common piece of calligraphy and clearly not intended for a wealthy or elevated audience. Instead, its 1,500 or so characters were designed to reach a less exalted readership, evidence of the social descent paper was making from the elite circles in which bamboo had been commonly used. But an earlier sutra in the collection, dated to AD 513, is altogether of a better quality – its paper softer, its calligraphy (in the 'broken wave' style) more mannered.

The best calligraphy on the Dunhuang sutras is set out with metronomic evenness, thanks in part to the column lines drawn with a thin brush down the page but also to the skill and care of the scribe. Such sutras, carried along the trading routes that tied China to Central Asia, popularized paper as nothing before them. They could not dislodge the bamboo Confucianism of high culture in China, or not at first, since this textual authority remained defining for the scholars and the court, as it did for their style of government. At the margins, however, there was more room for such innovations as paper and Buddhism. It was at those margins that Buddhism and paper rose together, woven into a new age of reading and writing across all the lands under the emperor.

Many of the Buddhist scriptures lying in the cave at the garrison town of Dunhuang had been written on the back of reports, contracts, ordinances or purchase orders; these account for more than four-fifths of the 40,000 scrolls and documents found in the cave. But the city was just one of the Silk Road trading towns which had received Buddhism from the west, just as it received silks, art, foods, medicine and clothes, as goods travelled between markets in the Indian subcontinent, China, Persia and the kingdoms of Central Asia, often on the backs of camels. (Buddhism probably arrived in Dunhuang from Central Asia around the first century AD.) In one painting at Dunhuang, a foreign itinerant storyteller stands in the foreground, a rucksack of scrolls strapped to his back. He may have been selling his scrolls and the cave's contents suggest they were likely to have been Buddhist scriptures.

Men like him were Buddhism's new couriers and the catalysts of China's first paper-made revolution. Contrary to its origins, China adopted Buddhism as a religion of the road and translators soon arrived in the country, a tiny group of linguists who transposed it to a Chinese market. It is odd that it should have worked at all. China's bookshelves had plenty of Confucian and Daoist classics as well as the quick-fire succession of commentaries that had followed them. Yet amid political turmoil and from a few wandering merchants and madcap translators, Buddhism found a way into China. First it adopted local gods and stories, many of them Daoist, then it began a textual tug-of-war with Daoism itself that was able to summon and then tap into a new market of religious readers. It even won emperors to its cause. Moreover, Buddhism did not simply arrive as a set of ideas for an intellectual elite to weigh up. Instead, it was delivered into a social setting, one it would both respond to and help to shape. Even its writings arrived not merely as claims to revelation, but as physical objects with social values and associations attached (or assumed). Indeed, it emerged in China (in the words of the classic 1950s study of Chinese Buddhism) as a 'way of life' – or even as various ways of life, given the varieties of social class and levels of cultural engagement among its many recipients or participants.[26]

The arrival of outside influence had begun on a road stretching far beyond territory with which China's rulers were familiar. In 138 BC Emperor Wu of the Han dynasty sent one of his palace gentlemen on a 2,000-mile journey to the west. Zhang Qian was a general from northern China and the emperor wanted him to travel beyond the lands of the western nomads, the Xiongnu, in order to make alliances against them.

Zhang left the Great China Plain by the Gansu Corridor, which arcs north-west to one of the driest regions on earth. From there, he passed through deserts and mountain ranges, and each day he expanded China's knowledge of foreign lands. On the way (and then also on the way back) he was captured by the Xiongnu. He penetrated as far as Ferghana and Bactria (parts of modern Tajikistan and Afghanistan) but he made no major alliances. Returning after more than a decade, he advised the emperor to build links with the countries to the west.

Zhang's mission laid the seed for China to open relations with Persia, and a trade relationship slowly took shape; the Silk Road was far from being the first trade route across Central Asia but, at least in terms of traffic, it was a network of highways where once there had been only streets and avenues. China now began to claim lands ever further along the Gansu Corridor to the north-west. Chinese silks started to arrive in the lands of the Oxus, Persia, India and, by the first century AD, in Rome. Chinese ambassadors were sent to the kingdoms and peoples of the west: Ferghana, the Yuezhi, Bactria, Tocharia, Parthia, Khotan and the Indian subcontinent. Dunhuang grew from a green village in the desert to a busy imperial command post and trading town.

Merchants began to travel the desert routes of western China from the west too, taking their camels, their wares and their cultures with them. Many came from the emerging Kushan empire of Central Asia, which stamped Buddhist icons onto its coins. (We largely know of the Kushan empire, which lasted from the first to third century AD and dominated Central Asia, from Chinese sources.) But it was carvings of warriors, princes and bodhisattvas which piqued Chinese interest in the Kushans. Their Buddhist play on the Greek ideal became a carved currency of the Silk Road; they used an adapted form of the Greek script and in their sculptures they fused Greek forms with Buddhist subjects. This Greco-Buddhist art travelled from its creators' Gandharan homes, not far from the headwaters of the Indus. It was here that sculptors first gave the Buddha human form.

The beginnings of Buddhism in the more populated east and centre of China survive only in myth and half-truths. The official histories of Chinese Buddhism say that in the first century AD, Emperor Ming had a dream of a giant golden man. When he recounted the dream, his adviser told him the man was the Buddha, so the emperor sent a group of ambassadors to fetch the sage's teachings from the peoples of Central Asia. They returned with the *Scripture in 42 Articles*, written on palm leaves, and a bundle of images. The scripture arrived on white horses a mile outside the capital's West Gate, so they built White Horse Temple on the site. (Here incense was burnt to honour the very scrolls of the Buddhist scriptures, which were said to give off their own light.) Emperor Ming's fact-finding embassy to the west is certainly plausible, given the growing trade links. Others have suggested Buddhism might have first come to China from the south, directly from what is now north-east India. One of the most improbable stories of Buddhism's early days in China is that some Sanskrit Buddhist scriptures were saved from the Qin's book-burning in the third century BC (although one seventh-century Chinese Buddhist reference work does mention the arrival of eighteen Buddhist missionaries in China in that century).

However the first contact was made, Buddhism's crucial entry into China by camel and pony from Central Asia, a few sutras and monks at a time, as trade and cultural curiosity grew in tandem. The heft of Dunhuang papers suggests it was these merchants who first delivered the new teachings. Interpreters at China's Bureau for Foreign Nations may have smoothed their way, but this message of salvation and enlightenment began its Chinese journey as an outsider.

In the Indian subcontinent, Buddhists had learnt to live in a society moulded by kings and brahmans, but their attitude to life had offered them an escape from it. In second-century China, where nobles lived in secluded luxury on large, landed estates and where the emperor was slowly losing control, the Buddhists' renunciation of material pleasures further highlighted the corruption and extravagance at court, and therefore had wide appeal. Shaven-headed monks preached escape from social obligations and pressures and offered the promise of life after death. Such messages held appeal for a broad audience, especially in light of growing political problems at the imperial court, where power had begun to splinter.

Thus, while palace eunuchs stoked factionalism in the capital, religious rebellions began to undermine imperial authority beyond the palace walls. In AD 184 peasants along the Yellow River rebelled, angered by acquisitive landlords, poor agricultural output and heavy taxes. Through secret Daoist societies they launched the Yellow Scarves Rebellion, so named after the yellow scarves the rebels wrapped, turban-like, around their heads. Confucians proposed answers to the malaise too, but their debates were too arcane, as they argued over chapter-and-verse trivia, constrained by the economical and sonorous poetry and prose they were accustomed to. Meanwhile, the energy of new ideas and fresh writing styles was increasingly caught on paper, not bamboo. As for Buddhism, it might have made its passage through the mountains and deserts of Central Asia, but that would make little difference if it simply remained an immigrant religion, marooned in its new home.

Instead Buddhism, like Christianity but unlike Confucianism or Islam, developed polyglot ambitions at a formative stage. The Buddha himself had told his followers to teach people the true path in their own languages. Nevertheless, the religion quickly appeared on talipat leaves – arriving in China, its greatest shift was from talipat leaf to paper page.

Buddhism was refreshingly non-elitist, a religion that spread through China on paper more than on bamboo, appealed to the lower and middle orders of society – partly thanks to its suspicion of unchanging social hierarchies – and adapted itself to local languages and dialects. In fact, Buddhist texts themselves were not always bought to be read. They might instead be sought out as charms, religious talismans or, in time, status symbols; thus, the illiterate bought them too. Illiteracy, which debarred the lower social orders from meaningful participation in religion, was not able to restrict Buddhism – or even its scriptures – to the reading classes.

None of China's historians of the time recorded this surprising social development. Their capital was cosmopolitan and, at the eastern terminus of the Silk Road, foreign faces were two a penny. But a handful of men from scattered lands arrived in the capital of Luoyang in the second century AD and began to translate the Buddha's words into Chinese. Among them were Parthians, Yuezhi and South Asians, and they translated from Pali, Prakrit, Sanskrit and Gandharan. One such translator was a Nestorian Christian who had settled in the capital in AD 148. Their translations were crude but they nevertheless induced a few intrigued Chinese scholars to join their ranks.

Between 60 and 220 there were fewer than fifteen translators of Buddhist scriptures in China whose records have survived. And yet these same men produced at least 409 Chinese titles. In the hundred years following the fall of the Han in 220, at least 744 Buddhist titles appeared, most of them translations (or purported translations).[27]

A tiny band of translators produced these works, men such as An Shigao from Parthia (in eastern Persia) and his friend Lokaksema from the Kushan empire. Both settled in Luoyang in the second century AD, driven perhaps by missionary zeal or else sent by patrons at home. An, a magician who claimed to be able to interpret the headlines (lightning, storms, thunder, earthquakes) of the natural world, translated some thirty-five Buddhist works into Chinese during the 150s and 160s. His friend Lokaksema, who translated at least fourteen scriptures, tried not only to translate meanings but sounds too, especially the sounds of names and technical terms.

By the 220s, there were translators of Buddhist texts operating as far away as Wuchang, hundreds of miles up the Yangzi River. Others travelled to Central Asia to gather texts. Among them was Zhu Zixing in the mid-third century. Perhaps China's first Buddhist ordinand, Zhu copied parts of the 25,000-verse Perfection of Wisdom Sutra, one of the first Buddhist examples of the treatment of writing as a holy practice. In one passage, the sutra explains that, if it is written down in a book and then held up to be worshipped, then neither men nor spirits can do unjustified harm to the worshipper. In 291, a translator called Moksala wrote a Chinese version and it became a central text of Chinese Buddhism.

Such men, largely ignored by official historians, penned a new literature which was a peculiar mishmash of foreign and Chinese terms. Through them, South Asian words poured into Chinese, enabling the birth of a more accessible literature, one aimed at an unusually broad range of readers and written in a more colloquial form. Their close attention to their target language even saw Chinese tones (of which there are four) defined for the first time. Their new, written Chinese religion was a linguistic hybrid: broad, syncretistic and impure.

In a context where scriptures were being constantly copied from bamboo onto paper, it is little wonder that Buddhism thrived on forgeries too. Usually purporting to be translated from a Sanskrit original, these bogus scriptures were simply trying to tap into the growing market for Buddhist writings. Few phenomena paint as clear a picture of a healthy market of paying customers as the emergence of forgeries.

Chinese Buddhism found an ideal partner in Chinese paper. It did, of course, spread on bamboo too, at least initially, but bamboo was a less neutral agent. Instead, it carried associations and these tended to be with high, classical Chinese. In terms of the time and cost involved in preparation, bamboo looked increasingly rarefied compared to paper. It often carried the classics on its surface, works whose language was beyond the comprehension of the vast majority of the population. Paper, on the other hand, was the ideal populist conduit for an imported religion. Both the style and the content of the writing could aim beyond China's elite – more people could afford to own its texts too.

As paper Buddhism spread into China, it developed a local character, adopting Chinese stories of sylphs and spirits, and borrowing from Confucian morality and ritual and from Daoist traditions of 'pure conversation' dialogue (a form of philosophical repartee). On occasions Buddha was even equated with the mythical old Queen Mother of the West (who was said to rule in the Kunlun Mountains of Chinese Turkestan), supposed saviour of the Han Chinese. Chinese Buddhism was a mongrel creation and its adaptability brought it enormous success. By the end of the third century AD, there were 180 Buddhist establishments split between China's two chief cities of Chang'an and Luoyang, together with some 3,700 clergy.

In paper, Buddhism had found its surface and in missionary-translators its proselytizers, but for ideas its chief source (beyond its own tradition) was Daoism. Imitating Daoism's methods helped to bring Buddhist writings into China's religious mainstream. First, Buddhism borrowed Daoism's special obsession with physical scriptures, a facet of Daoism that might have upset Laozi himself, who had argued for the hopeless inadequacy of language. Yet now, half a millennium after Laozi, Daoism's textual fixation was already acting as a prop to paper's fortunes.

Some Daoists argued that followers should sweep and purify any place where they found scriptures and set up an altar on the spot; there were many stories of mountains opening to those who knew the *dao* to reveal hidden scriptures. Daoist scriptures, above all the *Dao De Jing*, were the heat and light of primordial energy, like gods themselves, and encircled by spirits. One fourth-century Daoist scripture ordered its readers to bow to it before use, wash their hands and burn incense. A fifth-century Daoist courtier even made sacrifices to the scripture scrolls, and one of China's longest books of annals recounts a Daoist story entitled *Writing Red Characters on Paper then Swallowing It*.

In the third century AD, when China had divided into different kingdoms, a southern Chinese Daoist sorcerer (his family name was Li) was approached by the southern emperor for advice on waging war. Li asked only for paper and a brush. He drew some soldiers, horses and weapons on ten pieces of paper, tore them all up, then drew a big man on a fresh sheet, dug a hole in the earth and buried him. Having thus predicted loss and death, Li departed. (In fact, the emperor's last battle would be one he lost against the south-eastern kingdom of Wu.) Thus Daoism took bibliomania into the spiritual realm, investing written scripture with enormous power and authority.

In China, Buddhists took this love of books on for themselves, developing a new textual obsession. It was aided by the translation into Chinese of major Buddhist texts like the Lotus Sutra and Diamond Sutra, which described how devotees could win spiritual merit by copying out the Buddhist scriptures, even if the readership of these editions was tiny in number. In China these were especially popular texts; Aurel Stein alone removed more than 500 Diamond Sutras from the cave at Dunhuang. (He couldn't read Chinese.)

The Lotus Sutra argued that followers should regard a single roll of the sutra 'as if it were the Buddha himself'. In the many surviving Chinese stories of miracles brought about by the copying of scriptures, the word-by-word content of the scriptures is little more than a detail. Instead, their spiritual power is indivisible from their material, ink-and-paper presence, winning merit for their scribes, owners and worshippers alike. Copying a sutra was taken so seriously that a scribe would occasionally write it out in his own blood, in the hope it would work to the spiritual credit of a recently deceased parent (or himself).

It was not simply the heft of Daoist scriptures that Chinese Buddhists sought to emulate, but their content as well. The two religions began to imitate one another and in their scriptures and sutras they borrowed, plagiarized and finally fought in polemic. *Dharma*, the Buddhist path of virtue, became *dao*, Laozi's ineffable way, in the first translations. Daoism modelled its Heavenly Venerable Saviour on the Buddhist god Avalokitesvara, lifting his role, his character, his names and his image. Yet, like Chinese Buddhism, Daoism depicted the deity as female. Buddhism borrowed visions of salvation and the apocalypse from its new ally and both preached a future of demons carrying diseases and disaster.

Plagiarism became the means to create an aura of authenticity and respectability. *The Sutra of the Three Kitchens* was Buddhist, but Daoism flanked it with *The Scripture of the Five Kitchens*. Daoism's *Marvellous Scripture for Prolonging Life and Increasing the Account* spawned the Buddhist *Sutra to Increase the Account*. The two were close to identical. Some apocryphal writings found at Dunhuang formed a series that had its counterpart in a Daoist scripture serial, the order like for like. But with plagiarism came competition and the so-called 'Barbarians Conversion Argument', which would last for a millennium.

It began with the *Sutra of the Barbarians Conversion*, which intended to authenticate the rise of Buddhism in Chinese and to assert its superiority to all competitors. (Paul Pelliot, the French archaeologist and historian who arrived in Dunhuang after Stein, found a copy in scroll format at the caves.) Daoist writers responded to the sutra's claims by adopting mild xenophobia; they taught that Laozi had actually travelled west to convert the barbarian Indians to good manners, adapting his teaching to their lesser western intelligence. Daoism's *Scripture for Unbinding Curses, Revealed by the Lord Most High*

Lao, not only mimicked Buddhism's advice against witchcraft, but also dated Laozi's Indian journey to the ninth century BC, safely earlier than the birth of the Buddha himself. In turn, Buddhist sutras argued that the order had been the other way around – it was Laozi who was a disciple of the Buddha, Buddha had in fact come to China and Laozi himself was really from South Asia. Eventually, a referee was needed.

In AD 570 a copy of *Laughing at the Dao*, a list of Daoism's textual borrowings, was handed to the emperor by Buddhist complainants. At other times, the dispute became so heated that the emperor stepped in of his own accord; in 705, Emperor Zhongzong forbade Daoist temples to show paintings of Laozi converting the 'western barbarians'.

Although paper talismans, effigies and icons attached themselves to both religions, it was the alliance of brush, ink and paper which formed the currency of Buddhism's war with Daoism. As writing increased, monks themselves began to manufacture paper. In about 650, a novice monk near the capital was ordered to grow a large number of mulberry trees so as to produce the prized *xuan* paper and use it to copy sutras. Aristocrats funded huge monastic translation and copying projects as the steady flow of Indian originals and Chinese fakes continued to arrive or appear. In this way they hoped, like the Buddhist scribes they sponsored, to win merit in this life and the next. As production expanded, some monasteries even had to come up with new storage solutions to house their sutras

The Revolving Sutra Bookcase at Longxing Monastery sits in a square vermilion hall that flanks the central courtyard. There has been a monastery here since the sixth century. The twelfth-century bookcase looks like a cross between an old organ and the cross-beamed insides of a wooden ark, or a merry-go-round, with upturned eaves pointing out at intervals. It sits in a gulley, where men could push against wooden boards to rotate the structure, which stands twelve feet high. Sutras were placed between two hexagonal runners whose corners supported wooden pilasters. Several hundred paper scrolls could be safely stacked here. Moving the bookcase, and thus the sutras, was a means of winning merit, and all the more so because the movement was circular, which is itself holy to Buddhists. Paper scriptures were not merely to be read, chanted or owned – they were to be rotated too.

The war of bibliographies between Buddhism and Daoism saw each side trying to swamp the other with texts and to prove its worth to both the imperial elite and to wider society with the weight, authority and primacy of its scriptures. It was the rate of textual production that made this war so unprecedented. While writings on religion, politics and the cosmos had appeared on bamboo, silk and stone for centuries, they had been for a rarefied audience and often appeared gradually. Yet from around the fourth century, Buddhist and Daoist writers in China needed both to reach a broad audience and to do so as quickly as possible, since they were in clear competition for minds, hearts and market share. What followed was a religious cold war peculiarly suited to paper, the only writing surface which – thanks to cost, availability and ease for writing – could cater for their needs. For the first time, men were trying to mass-copy texts so that their writings could span the empire, filling the minds of its more literate subjects (and those they read and recited to) or else simply being sold as charms and talismans. Both sides canonized in earnest.

When purists complained of inaccuracies, output increased still further as scribes sought to produce more reliable versions. This increased the exchange of scriptures with other Buddhist lands. In 401 Kumarajiva, a monk from Kuche in modern China's north-west, arrived at the Chinese capital. Armed with a prodigious memory for sutras and knowledge of Chinese (rumoured to have been learnt in prison), Kumarajiva argued that Chinese Buddhism had been polluted by the language and beliefs of the Daoists. His arrival spawned a vogue for more accurate translations and commentaries.

As Kumarajiva travelled east, so the Chinese monk Fa Xian was travelling westwards, as far as Ghazni and Kandahar in Afghanistan, Palipatra in the centre of the Indian peninsula, and even to Sri Lanka. On his sojourns Fa acquired new texts, which were translated on his return. Fa ended up working alongside Kumarajiva too. Thus concern for accuracy bred still more manuscript exchange, more copying and more translations, and scholar-translators came to dominate Chinese Buddhism. The *Biographies of Eminent Monks*, a roll-call of China's Buddhist greats written in AD 530, listed 257 men and women. (The fact of listing women among the great and the good was a profoundly significant step, given the strict patriarchy of classical Chinese society.) Among the 257, 35 were translators, 101 were exegetes, 13 were experts in Buddhist law, 21 recited sutras and 11 were 'sutra masters'. Chinese Buddhism encouraged much wider reading than Confucianism had ever done, but it was a written religion above all, the practice of writing a scripture a means of accruing merit just as the possession of a written scripture was its own reward. Reading was often secondary.

Buddhism had begun on the edges of Chinese society. Thus, back in AD 279 the Jin dynasty of northern China owned only sixteen scrolls of Buddhist writings. (Each scroll is likely to have contained several hundred or several thousand Chinese characters, depending on the content, since many sheets of paper could be pasted together, end-to-end, in a single scroll.) Yet both Buddhism and its texts would become a feature of Chinese life and the paper scroll was a sufficiently convenient format to enable easy and widespread usage, especially since spiritual merit was accrued by reading entire texts end-to-end. (Reference reading of a scroll, on the other hand, was excessively time-consuming.) Indeed, the chanting of scriptures by monks became such a well-known sight and sound in China during this period that it gave birth to an idiom to describe the purring of a cat: the cat, so it went, was 'reading the sutras'. In time Chinese Buddhism even attracted imperial interest too, giving it a local customer with far deeper pockets.

So it was that, in the middle of the fifth century, the Song dynasty Imperial Library kept 438 Buddhist scrolls, representing hundreds of thousands of Chinese characters. By the early sixth century, when Ruan Xiaowu captured the books of the Liang dynasty in northern China, he was able to list 2,410 Buddhist and 425 Daoist titles. One of the first Chinese Buddhist catalogues, drawn up just a few years earlier, contained 2,162 works on 4,328 rolls. (A roll is a single sheet, whereas a scroll comprises two or more sheets pasted end-to-end.)

This expanding Buddhist canon was beyond the scope of any single reader, but still it continued to grow, adding to the weight of papers being produced across China. At the end of the sixth century, Emperor Wen (who may have converted to bolster his legitimacy) decreed that all Buddhist works should be copied and then deposited in the temples of large cities. A special copy of these works was made for the Imperial Library and more than 130,000 scrolls were transcribed over the course of his reign.

After the reunification of north and south China in 589, following three centuries of division, the emperor's patronage of Buddhism proved more significant still. An imperial catalogue of 597 lists 2,146 Buddhist works in 6,235 rolls. One historian of the day recorded there were ten or a hundred times as many private collections of Buddhist writings as collections of the Confucian classics; the official Sui dynasty (589–618) history went further, claiming there were thousands more copies of the Buddhist scriptures than of the Confucian classics. A parallel intellectual elite of monks and priests had appeared, one the Confucian elite had failed to take seriously enough. In their wake, Buddhist monasteries became large, wealthy landowners and reading and writing enjoyed a further flourishing. Buddhism had partnered with the written word to galvanize China for its own ends. What this foreign religion might never have achieved in China on bamboo, wood or stone alone – namely, a broad market of reader-buyers – had, thanks to the rise of paper, been delivered. Yet Buddhism had provided a breakthrough for paper too and, for the first time, there were hints of its more popular destiny.

The cave collection to which Wang Yuanlu had introduced Aurel Stein was dominated by Buddhist writings, but Stein found evidence of many other uses for paper as well. Some, like medicine labels, were pedestrian. Others, like personal letters and poetry, were the product of more elevated authors. Between them, the range of uses indicates paper's ability to seep into myriad different social circles and to be turned to very different ends: literary, medical, religious and more.

Drink features too. One of the documents found at Dunhuang was a ninth- or tenth-century poem called 'Argument Between Tea and Wine', in which the two drinks argue it out over which of them is superior. (In the end, water steps in to end the argument, pointing out that neither of them amounts to anything without water.) Another is a 'model letter' in which the author provides a series of sample letters to suit every occasion, with all the appropriate forms of address, sign-offs and turns of phrase written in.

The scribe records the date of his work: 13 October 856 (when translated into the Gregorian calendar). Formulas were used for regular forms of correspondence, not least because the Confucian hierarchy and its habits were revived with the start of the Tang dynasty in the seventh century. The letter appears on grey-white paper and it feels like a sweet bag. It is thick and, since it is not burnished, coarse too. This was a practical, low-budget product, reflecting a more mainstream market than the texts of Confucianism had ever known. Among the samples the author provides is one called 'A Letter Regarding Wine'. It has been evocatively translated by Lionel Giles, who first catalogued the documents Stein sent to the British Museum in the early twentieth century.

> Yesterday, having drunk too much, I was so intoxicated as to pass all bounds; but none of the rude and coarse language I used was uttered in a conscious state. The next morning, after hearing others speak on the subject, I realised what had happened, whereupon I was overwhelmed with confusion and ready to sink into the earth in shame. It was due to a vessel of small capacity being filled for the nonce too full. I humbly trust that you in your wise benevolence will not condemn me for my transgression. Soon I will come to apologise in person, but meanwhile I beg to send this written communication for your kind inspection. Leaving much unsaid, I am yours respectfully . . .[28]

The fallout from uncontrolled drinking might be dealt with in a few contrite and well-penned lines, but most of the scrolls were dozens of feet long. Scrolls were convenient provided the material ran to five, ten or perhaps even twenty feet. But forty feet was common too and several scrolls were longer still. Looking up later chapters could be painstakingly slow as you or your servant wound and unwound your way through the scroll.

One of the key developments represented within the Dunhuang collection was therefore the move towards a codex, the folded-leaf book we still use today, with its binding down one side. (The codex had already been used in the Middle East for centuries, the preferred format for Christian writings, but generally those were on parchment, since papyrus does not fold well.) Various different bindings were used to hold these books together, including *concertina*, in which you pulled the book open like an accordion, the misnamed *butterfly*, which pasted double pages to one another near the fold, and others besides, from *stitched* to *whirlwind* to *wrapped-back*. It was possibly the South Asian method of binding palm-leaf books that gave Chinese scribes and papermakers and shopkeepers the information they needed to create bound books.

These binding techniques are only apparent in some of the latest items in the Stein collection. Paper changed quality too, thanks in part to the Tang dynasty of 618–907, during which cultural production in general increased so markedly that rising paper demand pushed production efficiency up and prices down at the expense of craftsmanship. But paper's dip in quality was also down to the separation of the north-west from central China in the tenth and eleventh centuries, which squeezed supplies of China's finest mulberry paper to the whole Dunhuang region in the north-west.

The papers Stein saw and studied were made from hemp or the bark of mulberry or paper mulberry trees, with a few of them from ramie nettles. The older papers were usually thin, carefully finished and cut, and stained yellow or brown to prevent insect damage. Later papers were coarser. Some ten to twenty-eight of the sheets, each around a foot wide and two feet long, were stuck together to form long scrolls. We do not know why they ended up in the cave, although lots of explanations have been offered, such as stowing them away during monastery refurbishments, fear of what soldiers might do, or merely a desire to preserve ancient scriptures from the usual ravages of politics and repeated use. The region comprised historical borderlands and so had been subject to the attacks of various armed imperial claimants. For some of the ninth century it even formed the northern boundary of the Tibetan empire. Thus instability may have pushed preservation underground.

The trouble taken to preserve them is striking. It was said in Chinese that you should (literally) 'Respect-Cherish-Writing-Paper' (*jingzhi zizhi*), which you could paraphrase as either 'respect and cherish writing and paper' or 'respect and cherish writing on paper'. Paper was not easily thrown away if it had characters on it. (The historian Geoffrey Wood has pointed out that many of the more heavily used sutra scrolls found in the cave at Dunhuang had been mended with paper patches, an indication of reverential care.)

Such was the value placed on Chinese characters. This attitude may not explain the planned preservation of the cave's extraordinary contents, but it does reveal a mindset which, when threatened by wars and chaos, sought to protect and store writings in a cave, sealed off for future generations. Of course, it was wars and social unrest which had helped to dislodge the authority of the bamboo elite in the first place, providing a fresh material on which new writing and new ideas might flourish. Fu Xian, a third-century poet, recognized the possibilities that new material provided for new forms of writing. The earliest mention of paper replacing bamboo comes in Fu's *Paper Ode*:

> When the world is simple, embellishment has its uses. The rites and materials change as time goes by. As for making records, carvings replaced rope knots and bamboo was replaced by paper. As a material, paper is fine and worth cherishing, for its shape is square, its colour is pure and its nature is simple. Articles and expressive words are carried on its surface. It unfolds when I want to read it and I can fold it up again when I am finished. If you are living away from your family and relations, you can quickly write a letter and send it with messengers. No matter how distant your hearer, your thoughts can be expressed on a sheet of paper.[29]

Fu Xian had understood the advantages of paper. Bamboo was hard to carve into strips and silk was expensive, but paper gave unparalleled liberty to the author to write at length and in his own form. At dinner parties, guests would often compose quick poems on paper to one another, sparring in verse. In the 190s one such dinner guest, Mi Heng, was asked to compose a poem to the parrot that another guest had just given their host, Huang Yi. Huang was the son of the local warlord in today's Hubei province in central China. Mi's *Rhapsody on a Parrot* was quickly written but, thanks to easy reproduction on paper, widely admired.

Purple feet, cinnabar beak,
Green jacket, kingfisher collar.
Elegant-looking, many-colored;
Fine-sounding: 'Caw! Caw!'[30]

It was just such a game that had taken place at the summer party of the calligrapher Wang Xizhi when he wrote out *Preface to the Orchid Pavilion*. Once writing was liberated from the old bamboo confines, it found a new expressiveness. Lu Ji, an author and literary critic from southern China who lived from 263 to 303, wrote that poetry was 'born out of emotion and rich and colourful words'. He could never have made such a claim before paper's emergence.

Thanks to this new freedom, some who might have been too fearful or too poor to become an author in the bamboo age were able to pick up their brushes and to write. Fu Xian himself wrote that, thanks to paper, 'a humble person living in a backwater could become famous'. New forms of writing were born just as more men and women felt justified in setting their thoughts, ideas and experiences down in ink. Letters, diaries and written reminiscences became more common. Paper allowed both a broader spectrum of authors and a greater range of subject matter and forms. Writing spread from the court to the provinces too, following the migrating Han Chinese to southern China and so fanning out from the old capitals to capture new land-scapes in prose.

Many of the greatest authors of the chaotic period which followed the end of the Han dynasty in AD 220 were exploring deeper questions than Confucianism had generally addressed. Their approach became known as 'Dark Studies' and involved studying the cosmos as well as human language and society, offering an alternative to the strict moral hierarchy of the Confucian cosmos.[31] Dark Studies enjoyed its first flowering in the mid-third century and led writers to explore the origin of meaning and the limits of human understanding. New ideas appeared from such men as Wang Bi, one of Dark Studies' leading lights, who argued that nothingness was the true basis of everything.

One of the blossoming uses of writing was a form of literary criticism: the commentary. It was common practice to use smaller bamboo strips for commentaries and larger pieces for the classics themselves, but new forms had emerged since, some much better suited to paper. In the 280s, the literary critic Lu Ji even wrote in praise of literature's versatility:

> Words on a tablet state facts in elegant form, whereas eulogies are gentle and emotional. Inscriptions thrive off brevity ... Odes must be accessible and rich but essays should be complex and fluent. State reports need directness and grace whereas argument prefers explanation and exaggeration.[32]

Thirty years later, a scholar called Zhi Yu wrote *Criticism on the Literary Genres* to help identify the boundaries between different genres. In the book, he even pointed to the importance of paper in this process, arguing that the change in literary classifications was linked to the change in writing surface, which freed writers to express themselves less tersely.[33] What he recounted was a liberation of the sentence. More writing space allowed both more expressiveness and more exactness. Paper was not the only factor behind these developments; in political and social elites, for example, increasing non-Han influences (following the fall of the Han dynasty in AD 220) encouraged fresh forms of writing, too. But paper was a key factor, playing its part in the birth and popularization of new styles of writing.

The Chinese scholar Zha Pingqiu, writing in 2007, described how the switch from bamboo to paper brought a blossoming of both letter-writing and literary criticism.[34] Thus the great book-lover Ru Xiaoxu compiled numerous bibliographies before his death in 536, and wrote of the rapid growth of the imperial libraries during the Wei and Jin, barbarian dynasties which together lasted from 220 to 420. A new literary category was even created for individual poems and articles, a reflection of the growing importance of the individual voice in authorship. This new category also ensured that more intellectuals took such writings seriously. In time, paper even provided teachers with the option of relying less on oral instruction or their students' memories, providing scope for more information and more cross-referencing in the classroom.

Meanwhile, the many new works were gathered into imperial and private libraries. At the beginning of the Wei dynasty, the imperial librarian Wang Xiang compiled a book called *The Emperor's Reader*, a bibliographical work which ran to more than forty sections and contained eight million characters.[35] Such output had been unimaginable in the bamboo age, in large part due to its bulkiness, which hampered easy storage. Paper, however, was not yet cheap and, despite its increasingly widespread use, its cost could still jump quickly. The official annals, the *Book of Jin*, related that, in the late third century, the poet Zuo Si had spent ten years composing 'Rhapsody on the Three Capitals'. It was so widely loved – and copied – that the price of paper in Luoyang reached a new high. Years later, the Qing poet Yuan Mei (1716–97) wrote to a friend that the only people to have profited from its success had been the paper merchants.[36]

It was on the basis of Zuo Si's experience that the phrase 'the price of paper is high in Luoyang' emerged as an idiomatic way of praising a work of literature; it is still in use in China today. As individuals like Zuo Si wrote their own works, forging a path for more independent authorial voices, more people began to read works on paper. Literary forms once considered unsuitable were brought into the open. Short stories, love poems and folk tales, which were often oral forms in the bamboo age and not considered worthy of such an expensive surface, grew in popularity and in status. Before the advent of paper, books in China were, as the Sinologist Endymion Wilkinson neatly put it, 'a weighty matter'.[37] But the arrival of paper scrolls meant that books, for the first time, were highly portable; they could therefore be more widely owned, distributed and created (or recreated) too.

A growing number of people transcribed bamboo books onto paper, causing the production of forgeries and inaccurate versions to increase rapidly; forgeries especially increased in religious texts, as we have seen, but wrongly attributed versions of the Confucian classics from the Han dynasty survive as well. The new paper format lacked a standard, easily recognizable design and layout and was therefore an easy target for forgers. Yet these are the problems of a newly bookish culture, one able to produce reading matter at speed and at a lower price. Access alone had once been the greatest problem for readers. The rise of plagiarism, however, showed how far China had come.

Meanwhile, the Middle East was making a gradual switch from

papyrus to parchment, from a surface whose principal constituent was largely grown in the Nile Delta and Sicily to one which, though more expensive, could be made almost anywhere since it needed only the skins of animals. Parchment was named after Pergamum, now a ruin in western Turkey, the city where, according to Pliny the Elder, it was first widely manufactured. Parchment had been used by the Greeks for centuries but it only came into common and widespread use around the Mediterranean in the late third century AD. (Augustine of Hippo complained that the shortage of papyrus meant he had to use parchment instead.) The oldest complete New Testaments are fourth-century parchment codices. In most of Europe, papyrus documents are fortunate if they last two centuries before disintegrating. Even in Egypt, where aridity aids survival, only a few New Testament papyrus scraps survive.

China's ability to preserve and disseminate texts was unparalleled around the world. Thus official histories of the third to seventh centuries mention several moments of rapid inflation in the price of paper, such as when monks and nuns in their legions begin to copy out the work of one especially popular monk, or when an adviser recommends a ban on the copying of Buddhist scriptures to protect the price of paper and brushes. When the Daoist canon was finally compiled in the eighth century, it needed 3,400 scrolls. Laozi's own 5,000-word *Dao De Jing* was engraved on stone, in full, several times in the eighth and ninth centuries. In the seventh century, a monk called Huisi began to carve Buddhist sutras in stone and the canon he began was finished five centuries later, by which time it covered both faces on 7,137 steles. (A printed book version would run to half a million pages.) In 2009, in the hills south-west of Beijing, I visited a series of caves where thousands of such Buddhist stone sutras are still stored. Outside the caves, overlooking a valley far below, was a grassy stretch called 'Sutra-drying Platform'. Here, for several centuries, paper sutras were shaken out and aired to remove any accumulated moisture. Paper religions had diffused the written word as never before.

Political disintegration and war had helped to launch paper as the battlefield on which to fight out the course of China's future. In 403 a warlord became emperor of the Jin dynasty in northern China. Within a year, he had yielded his throne and fled westwards. But before that he made a decree.[38]

> In the ancient times there was no paper, and so bamboo slips were used; the use of bamboo slips had nothing to do with showing respect. From now on, all bamboo slips should be replaced with yellow paper.[39]

By then, however, most of them already had been.

6

Paper Rain

'. . . and what is the use of a book,' thought Alice, *'without pictures or conversations?'*
        Lewis Carroll, Alice's Adventures in Wonderland

China's first inventors were its first rulers. In the mythologies, they invented agriculture, established families and subdued the primal chaos. They were farmers, not nomads, and as they ploughed and settled, so they forged a patterned way of life. The Chinese character for 'civilization', pronounced *wen*, is at least 3,000 years old, and in its oldest usage it meant 'order' or 'pattern':

From the eighth century BC, that order began to drain away, spurred by political disintegration. The political centre failed to hold, and solutions were sought out to address the malaise. Confucius and Laozi were just two of those who sought to use their texts to restore civilization. Where some had argued simply for military might, *wen* now emerged as an alternative way to order the world. Indeed, written solutions such as theirs would redefine the very word for civilization. Thus the character *wen* soon came to mean 'literature' too.

Men like Confucius and Laozi wrote in high Chinese and most words were monosyllabic. Their sentences were spare and, as we have

seen, they used bamboo, wood or silk – materials that favoured conci-
sion. These were hardly the everyday utterances of peasants working
in the fields, yet as the spoken language developed in the centuries
that followed, the written word grew still further removed from daily
speech. This elite form became the written language of savants across
China, Korea, Japan, Vietnam and beyond. But its failure to evolve
alongside spoken Chinese would ultimately prove to be a weakness.

After the Han order fell apart in the early third century AD, nomads
ruled over northern China. They picked away at the traditional political order and they saw literary Chinese and its Confucian doctors as
part of the same weave. These nomad rulers may have sensed an
opportunity to break with the Han past in the form of Buddhist
sutras, since these texts began to arrive in numbers in the second century. As a religion of wanderers and outsiders, Buddhism carried
alternative visions of paradise to replace Confucius's failed earthly
utopia. Aristocrats, frightened at events on the ground, found solace
in visions of the afterlife, while peasants saw a new Buddhist salvation.

The new faith was spoken into being and its texts were recordings
of human speech rather than repetitions of archaic written forms. Set
out on Chinese papers, Buddhism continued to celebrate this spoken
heritage: its words were polysyllabic and its grammar was often colloquial. It endorsed popular speech by setting it down in ink. It even
inherited grammar and words from Sanskrit. The foreign monks who
brought these scriptures to China did not know much of the Chinese
script, still less hold it in high regard.

This new written style was still not simply a spoken language in
ink, but it chose characters for phonetic reasons far more than the
authors and editors of the clotted texts of Confucianism, which had
not moved on with the spoken language. Thanks to Buddhism's paper
imports, the written language was now capturing sounds spoken at
the time of writing. Monks sang psalms from manuscripts and they
were often the songs of the dispossessed and the outsider, of those
who were accorded only the lowest of statuses under the Confucian
hierarchy and patriarchy. Arcane and elevated scripts were not the
currency of widows and orphans; it was often such groups who
flocked to Buddhism's new lingua franca.

This switch towards the vernacular on the page did not merely make reading more accessible further down China's social scale. It also embedded both Buddhism and paper in lands across East Asia, beginning with Korea, Japan and Vietnam. This eastern odyssey of paper, fuelled by Buddhism and populism alike, is a crucial step in paper's global ascent. It is a story of new East Asian dynasties, cultures and invaders absorbing the practice of writing on paper for themselves, from Tibet to Mongolia to Vietnam. In the process, these cultures discovered paper's ability to cement political identities through setting down their languages – and thus, often, their identities – more permanently. Paper's role is easily obscured by the more obvious agents of change it partnered with: writing, Buddhism, the scroll trade, cultural exchange, the Chinese classics and the invention of new, local scripts, whether in the countries beyond China's borders or under the non-Han dynasties that would rule China itself (the Xianbei dynasty in the fifth century, the Liao in the tenth to twelfth, the Jin which followed it, the Yuan in the thirteenth and fourteenth, and the Manchu from the seventeenth to the twentieth).

Yet paper was the silent facilitator for them all. Even back in the first few centuries AD, Chinese classics on bamboo strips were decidedly Han. That made them unlikely sponsors for the strengthening of non-Han identities. Buddhist paper sutras, on the other hand, used a spoken idiom, borrowed words from a number of languages and appeared on a cheaper surface. Across East Asia, vernacular scripts were forged or else chiselled out of a Chinese original. Paper was the spur to this movement, in partnership with its new ally, Buddhism. Had paper simply provided a way to parrot Chinese culture abroad, the shift might have been unremarkable.

Instead, paper's travels through East Asia signalled a change. These were still lands within China's cultural orbit (in fact, in certain periods Korea, Japan and Vietnam were viewed as part of 'Greater China' – or at least as vassal kingdoms – by the Chinese themselves), but their adoption of papermaking from China did not lead them to imitate all of China's writing practices. As they received the technology of papermaking from China, they used it in different ways, a tendency that would only enhance their own, independent identities. Sometimes the physical form of the text changed, whether the quality of the paper or the layout of the text on the page. Far more importantly, the content changed too. Imported writings were Chinese, but native writings now took off as a result of China's influence. Paper was no longer simply a Chinese product for Chinese purposes and ideas. Instead, it was discovering a far grander destiny.

It all began in Korea, where Chinese influence was especially strong. In 1931, archaeologists excavated a Korean tomb from the Naknang period (108 BC–AD 313). Inside, they found a piece of paper, the oldest yet discovered on the Korean peninsula. It seems likely, therefore, that papermaking had begun on the peninsula by the early fourth century AD. Such an early date is not surprising: in 108 BC China's Han dynasty armies had conquered the capital of the dynasty that ruled northern Korea. Four military command centres were set up as a result, one of them remaining in Chinese hands until the fourth century. As a result, trade and cultural influence washed across the border.

Korean papers were made from hemp, rattan, mulberry, bamboo, straw, seaweed or paper mulberry. The moulds were made from bamboo or from Korean grass. Manufacturers invented coarser versions for raincoats, curtains and book-binding mounts. Sometimes they pressed several sheets together and oiled them to become floor mats. Single sheets were used on window panes and a tougher variety was used to make tents.

Ondol paper, formed by sticking several sheets of mulberry tree paper together and soaking them in rapeseed or sesame oil, is still used on the floors of traditional Korean homes, which are heated by an oven under the floor, from which smoke passes through a network of small tunnels. Ondol (which means 'warm stone') prevents water from rising through the floor and lasts for decades. The best Korean paper was Hanji, made from the inner bark of paper mulberry trees. It is still used for lanterns, for model flowers and as a writing surface.

The chief spur to papermaking in Korea was Buddhism, which also arrived from China. The results of this transfer of technology were seismic, perhaps greatest among them Korea's eventual development of its own script. This script would be, as we shall see, an expression and crucible of Korean identity. Papermaking became seminal to the development of a culture of writing right down the peninsula; when

Korea was reunified in the tenth century, paper writings were already the tools of bureaucracy and culture.

In 108 BC, Emperor Ming of China had sent troops to the peninsula to build command posts in the north. Chinese immigrants followed, doubtless bringing with them Chinese texts – texts set out on bamboo or wood, perhaps even a couple on silk. Both the Chinese language and its script had spread down the peninsula by the first century AD. By AD 75 local armies had defeated three of China's four command posts (the other held on until 313) but China left an indelible mark. There are early peninsular examples of writings inscribed on stone slabs. But the turning point came in 372, when the ruler of Former Qin, one of the many states of what had become a divided China, sent a Buddhist monk to the Korean peninsula.

Sundo, the missionary who arrived in the northern kingdom of Goguryeo in 372, brought icons and paper sutras with him. There may have been a few papermakers operating at the margins of peninsular society but Sundo provided the impetus for Goguryeo to steer papermaking into the mainstream. As well as his gifts, Sundo brought news from the ruler of Former Qin that Buddhism could protect the state from foreign armies and local rebellions.

As a result, Goguryeo adopted the syncretistic Buddhism prevalent in northern China, one already domesticated and nationalized by China itself. To it were added the patchwork sutras and magic of a Chinese occultist called Fo Tu Deng. The old Korean gods and spirits found new homes in the Buddhist pantheon too. The Chinese monk Tan Shi arrived in Goguryeo at the end of the fourth century carrying a large number of Buddhist texts. More followed, and in both directions too: Sungnang, a senior Korean monk, made the journey from Goguryeo to Dunhuang, more than 1,500 miles, in the early sixth century, simply to read writings from the Three Treatises school of Buddhism.

Buddhism soon prospered in the south-west of the peninsula too, with the arrival of an Indian monk called Marananta in 384 in the kingdom of Baekje. The following year, ten Buddhist monasteries were built in the Baekje capital and monks took up residence inside. By the sixth century they, too, were travelling west in search of more sutras. One such monk, Kyomik, returned to Baekje from India in 526 with five versions of monastic law and the texts of one of the three 'baskets' of the Pali Canon, the oldest written Buddhist teachings. ('Baskets' is a reference to how they were stored; 'volumes' would be a modern equivalent.) In a team with twenty-eight other monks, he translated seventeen volumes of monastic law. After him, two other Baekje monks wrote thirty-six rolls of commentary on Buddhist monastic law. Baekje's monks began to travel more regularly to the Chinese capital, while Chinese painters, artisans and monks travelled in the opposite direction.

Both these kingdoms, Goguryeo and Baekje, endorsed the written culture of China and of Buddhism, but Korea's political future lay with neither of them. Instead, the unification of the peninsula would be achieved by Silla, the small, backward kingdom in the south-east corner of the promontory. Silla was the last of the peninsular states to adopt Buddhism officially, in 527, and it only began to keep written records in 545. (By this point both Goguryeo and Baekje were making official records on paper.) In an apparent deal with the king, a Buddhist monk and adviser called Ichadon agreed to die (on the pretext of building a temple without royal consent). He was executed in expectation of a miracle and news was spread that, after death, his blood had turned to milk and his head had flown to a mountaintop.

It was just the excuse that the king, who liked the Buddha's teachings but faced opposition to Buddhism from a number of his advisers, had needed to justify the conversion of the realm. Buddha became the king's double and in 551, at the first 'Assembly of 100 Seats', monks read out the magical Golden Light and Benevolent Kings sutras in support of their ruler. (Several of these political assemblies took place, created simply to read out and explain these sutras.)

Buddhism then washed into Silla like a deluge. In 565 a monk called Myonggwan returned to Silla with 1,700 rolls of scriptures. An imported bronze Buddha followed and, in 576, a party of Chinese and Indian monks arrived in Silla with Mahayana sutras and relics. Silla was like an apprentice being entrusted with masterpieces. Sutras were not simply to be read. Many were decorated with gold and accompanied by polychrome paintings of the Pure Land of Ultimate Bliss, a Buddhist heaven. Celestial garlands and parasols arrived from abroad too. Buddhist art, icons and ideas became a national heirloom which the people could own and enjoy. As for the aristocrats, many of them were drawn to the idea of a unified faith, one that held the promise of order, learning, beauty and civilization, to replace the hoary old myths and shamanist religions peppered across the peninsula.

Sol Chong, Korea's greatest teacher of the Chinese classics, was a Buddhist priest. His translator father had coined a Korean Buddhist dogma and he knew literary Chinese, like several other Buddhist masters. Sol is thought to have been involved in the penning of a written language better able to express what had become a native faith. He regularized the nascent *idu* and *gugyeol* scripts, the first systems for representing the Korean language in Chinese characters. Sol would not simply choose Chinese characters for their meaning but, in some cases, for their sounds. His script would serve the whole peninsula when it was finally unified under Silla in 668.

Buddhism had laid a path for paper but Confucian and Daoist texts were popular within Korea's aristocracy too. In 682 Silla established a National School to teach Confucian, Buddhist and Daoist texts. Reading classical Chinese literature became not just a Korean hobby but a focus for high culture and it lasted down the centuries. (Even as late as the nineteenth century, Korea's Chinese literature ran to 10,000 anthologies and 200,000 books.) Koreans believed that a man's knowledge of Chinese literature and his ability to compose it determined his true literacy. Silla had put paper at the service of the state.

Korea's preoccupation with paper books took root. Silla's rule came to an end in the eighth century, but under the Goryeo dynasty, which rose to power in 918, the culture of writing continued to prosper. Buddhism experienced a second Korean spring and codified laws were written down. Where Confucianism was Chinese and conservative, Buddhism could be transformed for a local audience; by the fourteenth century, Korean prepositions, conjunctions and auxiliary verbs had all appeared on the page, either taken from Chinese characters or reinvented as Korean characters instead, and some Korean texts now used punctuation marks. These written details were all enabled by paper's form, availability and cost.

When a new dynasty took power in 1392, it switched the focus away from Buddhism and towards neo-Confucianism, and so the culture of writing on paper continued to prosper for entirely different reasons, as Confucian texts were copied out, studied and commented on.

In the early fifteenth century, an Office of Papermaking was established in the capital, employing almost 200 papermakers, mould-makers, carpenters and others under three supervisors. Politicians saw the uses of popular reading too; a fifteenth-century Korean king even wrote that books must be read across the country if government is to be effective. (Doubtless the works of Confucianism were not far from his thoughts.) Such dreams of reading across whole kingdoms were features of the age of paper, unimaginable when bamboo and wood were the cheapest surfaces available. Paper, it was understood, had the ability to make texts accessible to all.

In the middle of the fifteenth century Sejong, king of the Choson dynasty (which took power on the peninsula after the end of Mongol rule in 1392), created a Korean alphabet called *hangul*, which remains one of the most efficient in the world, thanks to the clarity and simplicity of its letters. The alphabet was a crucial development; paper had already delivered cheaper texts to the peninsula and in a far more portable, user-friendly format. With the invention of Hangul, Koreans now had an alphabet that was not only far more efficient for the scribes (in terms of time spent writing) but also more accessible to the people, since it was so much easier to learn. Hangul's letters are gathered into a single block for each syllable and can be arranged horizontally or vertically, making it faster to read than the Roman alphabet. (It is possible they began as pictures of the shape of the mouth and position of the tongue when each word was pronounced.) Hangul's signs each fill an invisible square, as in classical Chinese, but they are simpler, more like a collection of nuts and bolts spread across the page or building blocks in a toddler's playpen than free-flowing characters.

Sejong wanted to produce local editions of neo-Confucian works he had ordered from China, and he used a Mongolian script, *phagspa*, as the starting point for his new script. No earlier Korean alphabets have been found; the fifteenth century was the peninsula's golden age of inquiry and invention.

National sentiment propelled the new alphabet, allowing it to escape the censure of the Confucian bureaucrats, who preferred to keep writing and reading as elite practices. Hangul was an accessible and practical alphabet, one which could be more easily harnessed to the needs of Buddhism and agriculture than Chinese characters, the traditional stock-in-trade of government bureaucracy and Confucianism. The Korean Buddhist canon now grew and grew as paper imports arrived from China: from 5,048 scrolls in 730 to 6,197 in 1027. In 1087, Korea published a Tripitaka (or Buddhist canon) 6,000 books long. As early as 946, the Korean kingdom of Goryeo set aside 50,000 sacks of rice to fund the promotion and spread of Buddhism. Every Buddhist temple in the land was ordered to build a sutra treasury.

The great majority of these texts were, of course, set out on paper. Buddhism had blazed the trail for the peninsula but a similar movement would take place in the second half of the nineteenth century, when missionary translations of the Bible into Korean became rallying cries for a peninsula facing threats from larger powers. Not for the first time, it was paper that acted as the conduit for writings that would reshape the peninsula.

Even as an undecorated writing surface, paper has many associations. Since bamboo in China was loved as the surface of high Confucianism and high Chinese, paper became associated with mass production, an imported religion, wider reading audiences and more everyday uses, such as medicine labels, as well as non-writing uses, such as toilet roll. Although its status rose in China through the centuries, it had entered a writing world where bamboo and silk were revered. China has rarely been viewed as the producer of the finest papers in the world.

Paper (and the know-how required to make it) had probably reached Japan via Korea by the early fifth century. The Japanese were mentioned in an official Chinese history as early as AD 57; the *Book of the Later Han* refers to them as the 'people of Wa', and recounts that they regularly came over to China to pay tribute to the Chinese emperor. Those visits continued and, before the end of the Han dynasty in the early third century, Chinese were making the journey to Japan too, where they discovered a people who, they wrote in that century, ate raw fish and used clapping as part of their temple worship. As in Korea, papermaking was taken on enthusiastically in Japan and became the conduit of much more localized forms of writing and expression. Likewise, it enabled a native script to develop.

The Japanese did not write exclusively on paper: scribes also used wooden tablets called *mokkan*. This practice could stem from China's use of bamboo strips, but *mokkan* were used for administrative or bureaucratic writings rather than for the great works of philosophy, prose or poetry. This may begin to answer why Japan, as no other country in any age, has been so deeply entranced by paper for centuries, producing it with such rare devotion and exquisite skill. It is a mistake to view the paper used for text (written or printed) as no more than a blank page whose sole influence is contained in the words it transmits. This lesson is especially apposite in the case of Japan, where form, quality and beauty have been a particular obsession for papermakers through the centuries.

In 1919, when there was a debate over what paper to use for the final copy of the Treaty of Versailles, it was Japanese paper that was chosen. Where the Chinese invented and used paper, the Japanese nurtured and adored it. For much of Japanese history, hand papermaking has been ubiquitous on the archipelago, even during the twentieth century. There is a shrine to the god Mizuha-Nome-no-Mikoto in Okamoto, in recognition of his supposed introduction of papermaking, by way of a vision, to Japan in ancient times. Japan's own papers have since travelled worldwide and been put to the most ambitious ends.

In late 1944 a scattered handful of Americans saw hot air balloons in the skies over their homes. Balloons grounded across the country, from the Aleutian Islands to Mexico to Michigan. Each was white and thirty feet wide, and where the basket might have been, there was only a pack of explosives. Six people died when a thirteen-year-old girl pulled one of the balloons from a tree close to Bly in Oregon State.

Nobody else was killed but the balloons were an enigma – paper rain without a known sender. In an article in *Newsweek*, a journalist speculated that the balloons could have been launched from enemy submarines off America's Pacific Coast. Others argued that Germans at POW camps in the US had aired them or that Japanese Americans might have released them from their American War Relocation Camps. Balloons continued to land in states across the country, from Texas in the south to British Columbia and the suburbs of Detroit, Michigan.

A few sandbags fell from some of the hydrogen balloons as they hovered over the country. The US Military Geology Unit took the bags away to discover the mineral composition of the sand. They successfully identified the beach that had supplied the sand. Like the paper the balloons were made from, it was Japanese. The roots of this strange attack go back to a major discovery made earlier that century.

In the 1920s a Japanese meteorologist had been standing close to Mount Fuji watching pilot balloons as they drifted into the sky. Their change in speed pointed him to a hitherto unknown fact of nature: air moves faster a few miles above the earth's surface. He had just discovered the jet stream.

Two decades later the Japanese Ninth Army Technical Research Laboratory ordered 10,000 paper hot air balloons. Papermakers across the islands of Japan spent months mixing fibres before producing, pressing and drying sheets for the government. Nimble-fingered teenage girls glued the papers together with 'devil's tongue paste' (an edible Japanese paste made from the Konjac plant), always three or four sheets thick. Theatres and sumo halls became assembly factories. But neither the papermakers nor the paste-mixers knew what it was all for.

In 1944, more than 9,000 hydrogen balloons were launched into the jet stream and across the ocean in the hope they might kill American civilians, create havoc in American cities, start forest fires and harm American infrastructure. Only a thousand reached the United States; they killed just six people. An entire national wartime production project, taking two years and costing more than $2 million, had brought harm to no more than a small church picnic. In 1945, Japan stopped production.

Washi, the paper used to build the balloons, has been made throughout Japan's paper age for books, calligraphy, letters, envelopes, bags, umbrellas, lanterns, sliding screens, clothing, toilet paper and even a cannon. It means, simply, 'Japanese paper' (although it is certainly not the only paper Japan produces). Japan took a Chinese tradition and technology and then honed, studied, improved and standardized it, passing it through a thick, slow filter of quality control. It is the only paper many curators will use for backing or supporting ancient manuscripts. At the British Library in London, the frayed edges of ancient manuscripts are attached (with wheat starch paste) to sandy-coloured Japanese *kozo* paper made from the bark of paper mulberry trees. Japan may not have invented paper, but it cultivated it as no other culture before or since.

Before papermakers or scribes had arrived from the mainland, the islanders of the Japanese archipelago banked their stories with the *kataribe*, their storytellers. These stories retold animist myths in a culture that, according to the Chinese *Book of Sui* (completed in the 630s), used notches in sticks and knots on strings to make records and send messages.

Japan's contact with China and Korea begins with stories of a gift of slaves and an invasion of Korea, both dated to the second century AD. By the fourth century, Korean artisans were emigrating to the Japanese archipelago. Japan bought Korean iron to clear the land and plough its fields. According to the Japanese histories, a monk called Atogi then arrived from the Korean kingdom of Baekje in 404, became tutor in the Chinese classics to the Japanese royal heir and in 406 sent for Wani, another monk from Baekje, to replace him.

These two Buddhist scholars opened the classics to the Japanese court and recruited the Japanese ruler for China's cultural empire. In the story, Wani arrived with an improbable ten-volume edition of the Confucian *Analects* and a copy of the *Thousand Character Classic*, a sixth-century AD Chinese poem. Both would have been written on paper. The Japanese called him *the scribe* and he was credited, across the East China Sea from its birthplace, as the missionary of written civilization.

Thus Confucius and Buddha arrived in Japan hand in hand. When China descended into war, beginning in the third century AD, Korea became Japan's tutor. In 552, the king of Baekje sent seven monks to teach the eight Chinese schools of Buddhism. In 588, a Buddhist painter and two Buddhist architects from Baekje arrived on the islands to build a monastery. Kwalluk, a monk who landed in Japan from Korea thirteen years later, brought with him a portable library of writings on astronomy, geography, calendrics and the art of invisibility. He is known as the founder of Japanese medicine.

Tamjing is the best-remembered emissary of all. The Korean kingdom of Goguryeo had sent monks to Japan before but Tamjing, who sailed to the southern Japanese city of Nara in the seventh century, is credited with introducing the five Confucian classics, ink-making, papermaking and brush-making to the islands, according to the eighth-century *Chronicles of Japan*. In Nara, a city under the thighs of the archipelago, Tamjing arrived at Flourishing Law Temple. Here he painted murals in the Golden Hall (possibly scenes of the Buddhist paradise) but they were soon lost to fire. The temple was rebuilt – it is now the oldest wooden building in the world – and the paradisiacal scenes that decorate its walls may well be the work of Tamjing's own students.

Paper Buddhism became the conduit for literacy to grow in Japan, according to the Chinese *Book of Sui*, an official history. In 513, the imperial government launched a national programme to teach Chinese culture and Buddhism. Exchange students carried books, sculptures and paintings back to Japan from the mainland. The state kept paper records of diplomacy and domestic politics. A third of seventh-century Japanese aristocratic families claimed mainland blood; they welcomed the immigrant culture and so retooled their own. But there were patrons still further up the social scale, and two among them were crucial to the spread of paper and Buddhism in Japan: Shotoku and Temmu.

Shotoku was an early seventh-century ruler who dreamt of what the Buddha might do for Japan and what Japan might do for the Buddha.

He was such a strong believer in the value of Buddhist writings that he penned his own commentary on the Lotus Sutra. Japan's oldest written guide to governance, the *Constitution in 17 Articles* of 604, is also attributed to him. As he dripped Buddhist teachings into a land still swimming with clan gods and Shintoism, so those teachings began to dilute, gaining local colour. The Shotoku who survives in old Japanese accounts introduced every Chinese custom Japan adopted before the year 1000, including the use of chopsticks, but Shotoku's chief legacy grew out of his pursuit of Confucian and Buddhist scrolls.

The second great patron in paper's Buddhist appropriation of Japan was Temmu, who came to Japan's Chrysanthemum Throne in 672 after ending a civil war. Temmu ordered regular readings of the Buddhist scriptures at his new temple of Kawaradera. The year before he died he ordered Buddhist chapels to be built in the houses of aristocrats across the islands and for paper sutras and images of the Buddha to be placed inside them so as to facilitate worship. Thanks in part to Temmu, Shotoku's state-led Buddhism became the cud of emperors, and Japan's paper culture expanded to fill libraries across the islands. The combination of Buddhism and imperial determination had ensured that paper was one of the pillars on which Japanese culture rested.

Moreover, the flowering of this written culture in sixth-century Japan led to increased organization of government, education and religion, thus setting paper writings ever-more firmly at the heart of Japanese politics and society. It also allowed Buddhism to flourish on the page and a copy of the complete Buddhist canon was completed in the emperor's own Kawaradera temple in 673.

This attempt to record learning – and store it on paper – grew during the first half of the eighth century, with the first library established in 702 and a state scriptorium in operation by the late 720s. The library amassed Confucian and Buddhist works as well as recording state history. In the ninth century, its staff included four papermakers, ten brush-makers, four ink-makers and twenty copyists: by the late ninth century it was producing 20,000 sheets of paper per year, and in the tenth it was sourcing its paper from forty-two of Japan's sixty-six provinces. As for the state scriptorium, it had more than a hundred employees. An enormous copying centre, it was home not only to scribes but to title-writers and proofers too. An experienced copyist could write seven sheets a day, amounting to around 3,000 characters.

Across the country, meanwhile, sutras were gathered and copied in more than 500 temples, aided by tax breaks. Sutras were even written to protect the state, like the Golden Light and Benevolent Kings sutras:

> The four . . . Deva kings, wardens of the world, promise their countless followers [spirits] to protect the kings who carefully listen to this sutra and make reverent offerings, receiving and guarding this sacred text.[40]

Texts grew not only in number but in standing too, influencing more monumental arts. Japanese monasteries and temples built scriptoria as the central piece of their architectural jigsaws. Sutra repositories at the monasteries were soon outgrown by their libraries. The tight supply of secular texts kept these calligrapher-copyists in work. Fuelling it all, however, was the produce of the archipelago's papermakers.

For most of paper's history, papermaking has been a local venture. Local ingredients, local water supplies and local seasons have all played their part in determining the timing of the process, as well as its productivity. In Japan December has traditionally been a good time for a farmer to fell his paper mulberry trees, by means of a knife near to the root. He might then cut the branches to create small bundles, which filled a stack of wooden steamers for two hours, their lids fastened with ropes, over a cauldron. He would then pick out specks and stains with a knife, loosen the fibres further and place the bast (the flesh of the inner bark) into a vat of water and size (made from roots of trees and plants like the hollyhock) – size prevents the paper from absorbing all the ink. The winter air keeps the size stickier, so it does not sink to the bottom of the vat.

At this point the papermaker would then take a mould with a gauze surface, the gauze perhaps made from wild grasses he had picked locally and held taut by a wooden deckle (or frame), which also helped to keep the slurry from escaping before it passed through the gauze. The papermaker clasped the deckle in order to flush a charge of the paper suspension from the vat over the gauze by tilting it towards him and then away from him. He then peeled the page from the mould with a stick and laid it on a pile of drying papers behind him.

When his stack of papers reached several hundred, he would press them under a wooden board weighed down with stones. After several hours he would peel the pressed sheets from the pile one by one, laying them onto ginkgo boards and leaving them to dry naturally. Usually he would gather a team of family and friends to help with the process for two weeks over the early winter. In the nineteenth century, tens of thousands of Japanese family farms made *washi* in much this way every winter.

More professional papermakers, on the other hand, experimented with different vegetable fibres and produced luxury sheets they had coloured or decorated. In the eleventh century, shops appeared that sold only recycled paper, usually stained grey with inks. While Chinese papermakers made a priority of cost and efficiency, Japanese paper became the finest available. In the fifteenth century, a papermakers' guild was formed and, by the seventeenth century, there were 121 papermakers based in Kyoto prefecture alone.

Through the second half of the first millennium and beyond, the rise in demand for sutras, Confucian classics and the papers of government bureaucracy fed the papermaking industry across the Japanese archipelago. In some cultures, paper has been largely limited to functional uses, from protecting windows to merely carrying words unadorned. In Japan, however, paper sutras became ornamental treasures in their own right. There was still the same appeal to magic as in the first sutras to be sold along China's stretch of the Silk Road, but in Japan the page was a space for aesthetics as much as for grammar and spirituality.

In the twelfth century Taira Kiyomori, a Samurai general and politician, dedicated three sutras to Hiroshima Temple to thank the Buddha for his family's wealth. In 1164, he handed them over to a temple on Miyojima Island, just outside the city. The Lotus Sutra was in a bronze box inlaid with a filigree of gold and silver dragons and clouds. The sutra runs to thirty-three scrolls, their pages also decorated with silver and gold. Painted images sit beneath the text and floral arabesques connect scenes. Zigzag walls and serpentine clouds flank columns of

Chinese characters. These were sutras to be seen, and to be seen with; possessions that conferred status and an aura of high culture on their owners.

It was a form of Buddhism that was visual, rich and tangible. (By the twelfth century, sutra decoration in Japan had become a practised skill.) There was a strong emphasis on calligraphy: in the eighth century three renowned scribes known as the Three Brushes began to read calligraphic style as a guide to the philosophy, morals and personality of its scribe. Paper and ink were the partners of the scribe's ideas and emotions, a study in thought incarnate. Calligraphy became so esteemed that it could even obscure the characters, just as paper patterns and paintings were already doing, filtering the text into a new union of word, image and fibre. A ninth-century Japanese monk saw the elements of writing in still grander terms, celebrating their power to possess all things:

> Mountains are brushes, the oceans are ink,
> Heaven and earth are the case preserving the sutras,
> Each stroke of a character contains everything in the universe.[41]

Robes, chants, incense and processions showcased Japan's fresh artistic eloquence and courtiers sponsored Buddhist paintings and sculptures. Amid this aesthetic onslaught, sutras became revered, not only as symbols of piety and learning but also as powerful relics and talismans – and sometimes as microcosms of the *dharma* realm, embodiments of the whole of nature. Antiquated and abstruse, they increasingly needed commentaries to unlock them although many were never read, instead being placed inside images of the Buddha, buried or even used as a means to converse with the gods. From the eighth century sutras were installed as warden gods in temples, and even viewed as the doppelgängers of their authors, especially when stories of the miracles performed by them began to spread. In such a context, textual inconsistencies became irrelevant as the scrolls themselves held spiritual power. It later became an offence to mistreat printed sutras, to place them on the floor or to sleep with your feet pointing in their direction. Armed with the correct writings on its surface, paper could hold material power.

*

Chinese characters, however, did not travel as well as papermaking. Chinese arranges and orders its monosyllabic words to determine a sentence's meaning, but Japanese uses complex postpositions (in English, the 'ward' in 'skyward' is a postposition) after a poly-syllabic verb instead. Japanese words leech into one another (they agglutinate), unlike Chinese words, and Chinese cannot express Japanese sounds accurately. They even come from different linguistic families.

Buddhist texts in China and Korea had introduced ways of using characters phonetically. When they were copied in Japan, scribes re-arranged some words and phrases to suit their readership. Aided by the growth of literacy in seventh-century Japan, soon many Japanese words were being represented by Chinese characters purely for their sounds. Japanese had a limited vocabulary and each word had a greater range of pronunciations than in Chinese. Hence several Chinese characters might represent just one Japanese word.

This method spawned Japan's first native script, Manyogana ('borrowed names'), in which Chinese characters were used for their sounds. It took its name from the *Manyoshu* or *Collection of a Thousand Leaves*. This anthology, which comprised more than 4,500 Japanese poems, opened with a poem written in 759 and worked back in time, ending with poetry written perhaps as early as the middle of the fourth century AD. Most poems are in Japanese (although some are in Chinese) and the anthology lists 530 poets in all, some of them peasants, others kings. The poems come from all regions, from the Tsushima Islands near Korea to the Pacific frontier. They appear in a germinal Japanese script and depict Japan in writ-ten verse for the first time, an image of a land that had come of age on paper.

Manyogana became known as 'man's writing' but it was cumber-some, a lolloping script unsuited to an age of widespread reading. As more words were borrowed, Japanese writers invented two sylla-baries, *hiragana* and *katakana*, and each used elements of Chinese characters to stand for whole syllables in Japanese words. Yet a full transition was not made and today's Japanese script remains a mix of Chinese characters and these two syllabaries.

The *hiragana* syllabary became known as 'women's writing', as poetry in Japan became dominated by female authors, who were excluded from studying Chinese. In 885 poetry contests even became part of the high life. Judges delivered anodyne, generous appraisals and poetry became festive, even desultory. Births and funerals were marked in verse and picnickers wrote poems about nature. Men vied for the emperor's attention with the thirty-one-syllable *waka* ('Japanese poem'). After lovers had spent the night together, a poem was expected.

The literati continued to compose these paper verses until the twentieth century but, during the ninth to the twelfth centuries, women dominated authorship. Their poems were free of Chinese writing and they spurred, more than anything, the evolution of a national script, just as Chinese characters were being fully domesticated. *Hiragana*, a script of women and children, boosted literacy and aided the writing of fictional stories, diaries, essays and poems, just as it abetted national identity.

For all its beauty, Japanese is perhaps the most convoluted and inefficient writing system in the world, yet it appears on the most carefully prepared paper available. Paper has had more important allies in its history, and not one of the ideas or religions that have best served paper in its global progress was born in Japan. Moreover, in traversing Japan, papermaking reached its eastern terminus.

And yet Japan's contribution to the story of paper is exceptional. Paper had been useful in China since the first century AD and it had been the surface of exquisite calligraphy and artistry in China before papermaking was even practised in Japan. Japan's innovation was to develop an obsession that delivered to paper new possibilities. Papermaking itself became an art form in Japan and the concept of the page as a portal of beauty found its most remarkable early expression in Japan too. It is an obsession that has survived the centuries, from *origami* (paper folding) to *sumingashi* (paper marbling), and from calligraphy to armed hot air balloons, drifting over the Pacific.

> *On croit qu'on va faire un voyage mais c'est plutôt le voyage qui vous fait, ou vous défait.* (A man sets out to make a journey but it is instead the journey which makes – or unmakes – him.)[42]

*

The paper trails of East Asia thread south into Annam, west into Tibet and Chinese Turkestan, and north into the steppe. What China had sowed then blossomed in states across the region, but there were limits to its spread. Hinduism preferred speech to writing – the oral tradition of the Vedas emphasizes their tones and melodies in recitation and the conduit of the human voice, as well as their worded content; even today, printed editions of the Vedas are considered mere shadows of their chanted versions. Talipat (or 'writing palm') leaves were used as the Indian subcontinent's surface for Buddhist and Jain texts from the fifth century BC. Although papermaking had spread to India by the seventh century (possibly via the use of diplomatic letters), it was the arrival of Islam, centuries later, that would normalize its use. The subcontinent's own written religions continued to use Talipat leaves right up to the end of the sixteenth century. Moreover, parts of South-East Asia used Talipat leaves too, encouraged by the influence of Buddhism from India. This was especially true in today's Myanmar, Laos, Cambodia and Thailand.

Vietnam, on the other hand, was as receptive to Chinese paper as it was to Chinese civilization. Largely under Chinese rule for a millennium from its absorption into China's empire in 111 BC, Vietnam took on Chinese characters in the third century AD and, with them, papermaking, although Buddhism (and the use of paper) was present at least a century earlier. In the Tang dynasty (618–907), China established Annam (now Vietnam) as a frontier province for non-Chinese peoples and absorbed the local ruling classes into its bureaucracy. When the Tang weakened, Annam began to loosen its ties, and by the eleventh century it had a centralized state with a ruler at Hanoi. The ruler, Dinh Bo Linh, used monks as his lever of control. Monks – and their writings – helped him to manage his relations with China, with labour, with public opinion and with the masters of the national wealth. Under his successor, Chinese political theory was remodelled to fit local tastes. A growing national identity spawned a Vietnamese script. Supposedly invented by a poet in the thirteenth century, the script was in fact far older and would help to enable Vietnam's independent identity.

To China's west, Central Asia remained too poor and fragmented to turn to papermaking in earnest. The break-up of the Kushan empire in the third century AD heralded the end of unity in the region, and until the arrival of Islam, Central Asia remained divided up between larger empires or simply local princedoms. Paper did at least enter the region in this period, however, given the dates of documents found at Gilgit in Pakistan (on sixth-century paper) and Mount Mug in Tajikistan (written on early eighth-century paper). China itself was disunited between 220 and 589 and often beset by war. Distance, deserts and political fractures may explain why papermaking never travelled west in these years. Central Asia already had scripts which appeared on other surfaces, so Chinese characters would not have held the same appeal as in, say, the Japanese archipelago, where Chinese characters arrived in an illiterate culture.

Tibet, on the other hand, began to make its own paper in the seventh century, when it received a good working script from India (although which one remains disputed), as monastery scribes copied old texts and added to them, penning a plateau Buddhism that would be popular as far away as the Mongol steppe. Tibetan monasteries produced clay charms called *tshatshas*. The name comes from a Sanskrit word that means 'copy' and this was the work of scribes too, those men and women employed to accrue merit through imitation of a holy image or text. Meanwhile, power began to shift from the local aristocracy to the Buddhist clergy and history began to be described and recorded by Tibet's Buddhist monks, who may also have eradicated many older texts. A fresh form of papermaking now spread across the Tibetan plateau, using the root of a Chinese stellera plant. The plant's root contains enough poison to repel moths, mice and bookworms like the death-watch beetle. Tibetan paper was durable, light and flexible and it cemented a literate Buddhist community across the plateau and beyond. Tibet's senior monks even played a role in domestic policy and foreign relations.

Tibet received new northern neighbours when the Uighur Turks conquered the Silk Road's Tarim Basin in the mid-ninth century, leaving their capital of Ordu Baliq in the steppe to the north. They already had a script but, arriving in what is now north-west China in the 840s, they acquired the practice of papermaking too. By this time the Uighurs were settlers but their unusual role in the story of paper was played out through their influence on a nomadic people who came from the same steppe region the Uighurs themselves had left a few centuries before. Although this took place several centuries after papermaking had spread westwards and out of East Asia, it is a development that belongs to the same, East Asian story. And so, for a moment, it is important to depart from the chronology, stepping forward into the twelfth century.

As you approach the old capital of Mongolia, a square of evenly spaced white mounds looms in the distance, like gulls perched on the poles of a jetty. These are the 108 stupas that surround the extinct capital of Karakorum. Inside, a few temples remain but the overall impression here is of vacancy in what was once a city of canvas. As you enter, you are on a path once trodden by pilgrims from around Asia and by official envoys from as far afield as Western Europe. Aside from a few monks chanting sutras and fingering a line of prayer flags, there is little sign of the glories of what was the capital in the golden age of the Mongols, seven centuries ago. Karakorum tells an Ozymandian tale.

The old capital of East Asia was not only a city of tents but of clerks too. In the thirteenth century, each minister needed a scribe who could write decrees in the various principal languages of the empire – Chinese, Tibetan, Uighur, Tangut, Persian and Mongolian. There were departments for sacrifices and shamans, for managing the merchants, for postal relay stations and for the emperor's treasuries and arsenals. All of them came under the supervision of the second most senior minister, a Nestorian Christian called Bulghai. Although senior officials were usually Mongol, the clerical staff tended to be from elsewhere and a third of Karakorum itself was set aside for these paper-pushers. In just half a century, the Mongols had grown from an illiterate tribe on the eastern steppe to the guardians of a world empire run almost entirely on paper, in several scripts and languages.

The change had come with a man called Temujin, who had spent much of his childhood as a fugitive trying to look after the family his late father had left behind. Temujin's education was in subterfuge, cunning and strategies of attack as he rustled horses and fled early sufferings to raise the yak-tail flag, the ruler's standard, at a giant meeting of the steppe tribes. This was the foundation of Mongolia and he would come to be known as Genghis Khan, Lord of the Oceans, in a land without a sea.

Genghis's ambitions took him to Korea, northern China and to most of Turkestan. He conquered the glistening city of Samarkand, released (according to a leading Persian historian of the day) cats and swallows, doused in oil and set alight, into the Persian city of Nishapur, preached that he was the scourge of God from the steps of Bukhara's Friday Mosque, mutilated the residents and destroyed the buildings of Beijing, and hounded down Shah Muhammad, his last rival in Central Asia, until the shah died of exhaustion on an island in the Caspian Sea. His dynasty conquered Persia, added southern China to its empire, launched a naval attack against Japan, subjugated Tibet, swept through the Middle East and reached the gates of Vienna. The new Pax Mongolica that spanned Asia in the thirteenth century allowed products and religions to travel further than ever.

But the travels of Temujin were a journey in reverse. When he reached the oasis towns of Chinese Turkestan, he found the Uighur people scattered across them ready to accept membership in the Mongolian empire. The Uighurs were settled and ordered, and their administrative skills impressed their new ruler. (A Chinese official had already persuaded him that his new Chinese subjects were more useful as taxpayers than as corpses.) But when Temujin stayed with the Uighurs, it was their alphabet that particularly struck him and he commissioned Uighur scribes to pen a variation of it for the Mongolian language too. This was their first script.

As the Mongol empire advanced, the Uighur traditions of chancery and scribes were grafted into Mongol governance. Communications were the great strength of Pax Mongolica, with post stations at gaps of a day's journey along the main trade routes, just twenty-five or thirty miles apart according to Marco Polo. Each post held stocks of horses and fodder so that a messenger with the appropriate tablet of authority could change horses immediately if he needed to. An express courier might cover more than 200 miles in a single day. But the written Mongol messages that zigzagged across Asia were only possible thanks to a handful of Uighur scribes.

The success of Temujin's new empire owed much to what it borrowed: writing from the Uighurs, a postal system from the Khitans, government structures from China and elsewhere. Temujin did not impose a Mongol way of life. Instead, he reformed that way of life and it was reformed above all by his quest for a Mongolian script. Armed with a script, he was able to ensure that the founding myth-history of the Mongolian people was written down on paper in his own language as *The Secret History of the Mongols* (although it initially appeared in Chinese). The myth of Mongolia was crucial to its survival, since the state itself was just an amalgam of different tribes, and the book survives today as testament to the power paper, ink and script enjoyed in forging the idea of the Mongolian people, who would conquer half of Eurasia.

Prior to the Mongol conquests, alien dynasties governing China from the early tenth century had adopted their own alphabets so that paper could keep and carry their own decrees, histories, poems, sagas and ideas. Under the twelfth-century Jin dynasty of northern China, when the country was ruled by the Jurchen people from Manchuria, the dynasty developed the ideal of the amateur gentleman-scholar, who could combine his literary training with dexterous use of the brush. A gentleman would never condescend to paint a mural; he would only write on silk or paper. Book printing (to which we shall return) expanded under the Jin too, resulting in inflation when the dynasty sought to print money to fund its deficits. When the Mongols attacked China in 1206, the Jin printed still more money to fund war and its currency again sank in value. After the conquest of China, the Mongol government kept some of its old hierarchies but never reverted completely to its steppe traditions. Instead, it allowed the scripted bureaucracy of China to become its umbrella of control over the country.

The late thirteenth-century Yuan emperor Kublai Khan, subject of Coleridge's poem, chose to move his capital from Karakorum to Beijing, to wear Chinese clothes and to adopt the Chinese way of life, its books and letters and administrative documents. The khan's move began a cultural outpouring that marked the rest of the Yuan dynasty, signalled by the founding of the Academy of the Pavilion of the Star of Literature in the first half of the fourteenth century, a centre designed to cultivate arts and literature. Toghun Temur, a later Yuan emperor, would spend hours in the academy practising his calligraphy and viewing his art collection.

The dynasty published and translated many books in the early fourteenth century, including a Song dynasty compendium, two works on the old Emperor Taizong, the huge *Comprehensive Mirror for Aid in Government*, the *Classic of History*, the *Extended Meaning of the Great Learning*, the Confucian *Book of Filial Piety*, *Biographies of Women*, studies of the *Spring and Autumn Annals* and an official Yuan dynasty work called *Essentials of Agriculture and Sericulture*. Confucian principles were laid down in books in the Mongol script to teach Mongolian bureaucrats how to run an empire whose administrative infrastructure was paper. As the Yuan dynasty settled, opera and drama thrived. Moreover, the Yuan dynasty's interest not just in elites but also in the common man paved the way for the beginnings of the Chinese proto-novel as well as for new themes in poetry and literature. The written arts were claiming new subjects.

The Mongols were the least likely of paper's subjects, and their adoption of paper for writing is not so much another chapter in the conversion of East Asia as a postscript to a later edition. They were centuries late in taking up paper and only a new way of life – as imperial settlers – could persuade them of its virtues. Yet persuade them it did. Even nomads could see the ability paper had to control and to legitimize, especially in the administration of an empire. In winning over the Mongols, paper had proved its extraordinary capacity to convert even the most alien powers to its use. Thus the Mongols act merely as an especially powerful illustration of what paper had in fact achieved across East Asia in the middle centuries of the first millennium AD. For it was during this earlier period that paper seeped into all the lands of 'Greater China', and so claimed half of Asia. Scripts attached themselves to local and ethnic identities and helped to set those identities down on paper. Texts multiplied and the technologies of characters, alphabets, brush, ink, bamboo-strips, paper, rubbings and woodblock-printing advanced and began to be shaped differently across East Asia and its borderlands. What had begun as an assertion of the Chinese way had been turned to new ends, from Buddhism to local independence movements. Paper was no longer simply a Chinese phenomenon. Its global quest had begun.

7

Papyrocracy

No poet in the world can ever have enjoyed greater contemp-
orary popularity as Po.
        Arthur Waley, Bai ['Po'] Juyi's twentieth-century translator
                                            and biographer. [43]

In the eighth century, Chang'an, China's imperial capital, was the most fabulous city in Asia, if not the world. Laid out like a chessboard, as one of its poets put it, the great city (now called Xi'an) was home to a million inhabitants and filled with the wealth, beauty, intrigue, commerce and clamour of one of the most cosmopolitan and cultured metropolises on earth. Only Constantinople and Baghdad could rival it.

China's divided north and south had been reunified in AD 581 and an age of prosperity had followed. At its heart was Chang'an, a city that teemed with merchants. Chang'an's markets lay in nine separate areas and the Eastern Market alone had 220 lanes, each one dedicated to its own product, such as meat, iron, clothes, steamed buns, bridles, saddles, fish or gold jewellery. Persian stores stocked semi-precious stones, precious metals, elephant tusks, gems, sacred relics and pearls. Rugs came from Bukhara and from Persia as tribute to the imperial court. One Tang empress acquired a couch made from ivory which sat in a tent covered with gold, silver, pearls and jade. Its covers were made of rhinoceros horn, its cushions of sable, its mosquito felt and its 'dragon whisker' or 'phoenix pinion' mats were woven with reeds.

The markets sold clothes stamped with multicoloured patterns. Stalls offered cloves for bad breath and 'barbarian body odour' treatments. Women could buy skin rubs, blackhead removers, face creams, cosmetics, skin gloss and beauty marks in the form of a bird or the moon. They plucked their eyebrows and painted in new ones with tinctures, applied rouge to their cheeks and cinnabar glosses to their lips. Women's cosmetics, ornaments and clothing were represented – such were their availability and social importance – by six gods, one each for ointments, eyebrow dyes, face powders, glosses, jewellery and gowns.

Public holidays were common and by the middle of the eighth century the government marked twenty-eight such days, as well as a further forty-eight sanctioned days of leave for officials. At festival circuses, acrobats walked in pyramids along a rope and fencers fought, while others juggled swords or pillows or pearls. There were pole acts in which an acrobat danced on a crossbeam attached to a sixty-foot-high pole carried on the head of another performer who walked around the stage. Daoist priests set up tall ladders for one of their number to climb up barefoot and brandish his sword against epidemics – a spiritual approach to mass vaccination. Others danced. Palace women and the new civil service graduates played a sport akin to football on the Cold Food Festival on the fifth day of April. One emperor destroyed the ancestral temple built for his predecessor's daughter to make way for a 'football' pitch. In China it was often military officials who played a game like football with a leather ball. Polo was, then as now, an elite sport, but in China each team had sixteen riders and a band usually performed during matches.

Amid all these glamorous displays, paper was ubiquitous. Gate gods were painted onto papers stuck to the lintels of doorways as protection against malign spirits. There were no glass panes for windows, but oiled paper worked well if silk was too expensive. Lanterns, made from light bamboo frames, were panelled with paper too. During the three-day Lantern Festival, the curfew was lifted and residents could walk with their lanterns under the full moon. Wealthy citizens spent to impress, and in 713 the emperor ordered a 200-foot-tall lantern to be stood outside one of the city gates, covered in brocades and silk gauze. Paper had begun to replace the traditional loo sticks (the sticks used to clean yourself after defecation), and when an area east of the capital was plundered in 767 by the troops of a returning military governor, so much booty was taken that local mandarins had to wear paper clothes.[44]

However, it was written papers which lay at the heart of cultural life in China's new golden age. Even in times of war, men of learning would draw on their readings to help them through the hardships, and they sought to protect important works of literature. This age of scholarship allowed high officials and literary enthusiasts to enjoy a way of life facilitated by paper, whether in public administration or privately written poems. (More than 40,000 poems, composed by more than 2,000 Tang poets, survive today – Ezra Pound, Father of the Imagist Movement in the twentieth century, would look to the simplicity of form and expression achieved by the Tang poets as a model to follow.) China, at peace after reunification in 581, had grown in prosperity, and bookmanship had blossomed as a result. (A reunified China also meant paper correspondence could travel much further.)

Paper (often manufactured in and around Chengdu in China's south-west) was now inexpensive enough for private owners to record more and more of their lives on its surface. For the first time it becomes possible to follow a private life closely on paper, or at least to follow the life of a man who inhabited a paper world. More than that: it makes it possible, centuries later, to access that most hidden of stories – the inner life of an individual. Of course, only so much access is possible and, even then, there are only a handful of individuals for whom the paper trail is long and deep enough. But it is extraordinary to read such personal insights and experiences of men who lived in China twelve centuries ago.

One man in particular stands out as worthy of just such a focus, a man whose life was lived out on paper through poetry, politics, religion, administration and letters. Thanks to his political vocation, his personal hobbies and friendships, his exceptional mind, his social concern and his poetry, his life displays, as few lives could, the ubiquity of paper in the lives of eighth-century China's educated elite. Yet his life also shows how paper was not simply used for functional forms of writing, such as government administration, exams and official correspondence. Instead, through his letters, poems and diaries, we are invited into his experience of life too, whether it is the pain of personal loss, the pleasure of intimate friendship, frustration over government corruption, anger at extremes of wealth inequality, spiritual epiphanies or moments of intense communion with nature. The sheer volume of his writings, the range of his subjects, the frequently intense expression of mood and the risks he sometimes took with political criticism have left us with something unprecedented in China before his time. Only paper could have delivered us such a legacy.

10 Text as process: *Northern Qi Scholars collating classic texts.* An exceptional example of early Chinese figurative painting, Yan Liben's eleventh-century work combines elegant forms with clumsy incident, capturing a moment of discord (the scholar centre-front suddenly preparing to leave, being grabbed by a fellow scholar who has dropped his instrument, and knocking a plate of food on to the platform) in what is otherwise a carefully composed harmony. It has often been thought to recall the moment when, in 556, Emperor Wenxuan invited twelve Confucian scholars (Yan depicted just five) to collate copies of the Chinese classics for the crown prince's education. But their drinking, their state of semi-undress and their casual posture suggests instead an independent gathering, free of official constraints, one more in tune with nature than politics. The scholar to the right is wearing a red, nomadic gown and sitting on a 'nomad seat'. Around him, the servants have various roles in the creation of the written work: one handles the paper, another prepares a new brush, a third holds a roll of spare paper, another reviews the writing and a maid brings an official girdle (symbol of rank) to the scholar-scribe. (Photograph copyright © 2014 Museum of Fine Arts, Boston)

Given the interests that would define his life, Bai Juyi was born sixty years too late, in 772, as wars and famine were unravelling China's golden age. China's registered population, numbering 50 million two decades earlier, had dropped to just 17 million by the year 764. A rebellion, famine and invasions to north and south had made the rest of the eighth century a struggle across China. Danger stalked the countryside and the government had to sell Buddhist ordination licences and civil service examination certificates to replenish its emptied coffers.

Bai's family were poor scholar-officials who lived in the old heartlands of the country. He and his brother were brought up by their maternal grandmother and Bai later claimed to be able to read a few characters when still a baby. At the age of five or six he was composing verses, and by nine, he later wrote, he had mastered rhymes and tones. Years afterwards, he wrote to his nieces and nephews to encourage them to pursue learning too.

> Life swindles those who cannot read;
> I am blessed with knowledge of brush and script.
> Life swindles those who lack position;
> I am blessed with government office.[45]

In 783 the emperor was forced out of the capital by rebel armies, but by the autumn of 784 his dynasty, the Tang, had been re-established. The following year, however, drought dried up the wells of the capital and the fields that supplied it with food.

The history of China could be written as a tale of waters. It begins with the gods separating the seas and the land, then shifts to the appearance of farming and irrigation, to Daoist geomancy (*feng shui*, literally 'wind-water'), to the Grand Canal that took southern produce north, to the problems of Beijing's desert-side dryness, to the communist hydro-dams of the twentieth century and the Pharaonic South–North Water Diversion Project of the twenty-first.

For Bai's family, the unrest brought on by the drought forged a division and they scattered. Bai was taken to live in cities in the Yangzi Delta to the south-east, a refugee fleeing northern unrest. His first dated poem, written on cheap paper, according to the last line, is from this time and it replays the old Chinese theme of absence.

Banished from my old home, what use my longing?
Chu River and the hills of Wu lie between us.
Yet today a kind friend helps me send my brothers
A few lines of wistful remembrance, brushed out on a scrap.

By 788 Bai was sending poems to Gu Kuang, whose verses were famous across China. But Bai was working too, probably as a clerk or copyist in a local office. In fact, his family was approaching poverty. Thus, when his father, a minor official, died in 794 Bai was forced to postpone taking the civil service exam. Besides, in keeping with Confucian tradition, he had to mourn for three long years before taking on any official work again.

In 799 he travelled to the eastern capital, Luoyang, to bring money to his mother, who was ill. But before crossing to the north bank of the Yangzi, Bai stopped at a small town to take his provincial civil service exam. The poems he brushed out proved both his knowledge of the classics and his skills as a wordsmith and he passed through to the final round in the capital, due the following spring. In 800, at the age of twenty-eight, Bai passed China's biggest exam in the top tier, his papers putting him in fourth place out of nineteen new Advanced Scholars.

The civil service examination had come of age under Empress Wu at the end of the seventh century. The examiners of the Ministry of Personnel grew powerful and the exams began to undermine class stasis in the capital. However, in 737 the Ministry of Rites took over and the power of the examiners declined. Senior officials in the Tang were not usually top exam graduates and, even among those who were, most had found an influential patron who could make their case to the examiners in person. Bai sent a fawning letter to the official in charge of the imperial tombs but nothing came of it. He must have excelled in at least the four disciplines that mattered most: manners, eloquence, calligraphy and knowledge of the law (though he claimed in his writings that he knew very little of calligraphy).

At the final stage, there was a choice of two exams, one classical and one literary, and candidates had to marry Confucian moral principles with the pragmatics of policy-making. The classics exam was a test of rote learning and of memory, requiring candidates to recount state-sanctioned interpretations of the classics. Bai chose the literary exam, the product of a new age of belles-lettres that had begun in the 690s. Among the new questions, many asked for answers written as eulogies, rhyme-prose or poetry. The most successful men in the empire were wordsmiths, and the exam they took was the talent-spotter for the most advanced form of government in the world. Bai happened to take his exam the year a reformer was attacking corruption in the exam system and he brushed his way to fame with answers on Confucius–Laozi paradoxes, the mechanics of nature and the proposed revival of harmonized purchase to stabilize the grain market. He wrote 'Parting from my fellow candidates' before his departure, a few weeks later, to mourn his grandmother's death in Luoyang.

> Full ten years, all my days in books;
> But now at least the honour's won.
> Success is not great prize to me,
> Pride comes when parents hear the news.
> Six or seven fellow students
> See me off from the capital
> . . .
> Flutes and strings send parting songs.
> . . .
> Wine smooths out the road ahead:
> Horse hooves move at lightning pace
> Through the spring day, and on to home.

When Bai turned twenty-nine (thirty by Chinese reckoning) he was unemployed and, as he morosely put it, thirty is almost a third of a hundred. His hope lay in the official employment test, writing his answers in a stylized form. Bai's answers bulged with allusions and old proverbs as he proposed solutions to a fistful of social disputes and injustices. He passed with seven others and became a collator in the palace library.

Bai was now a minor clerk and his job took up just two days of each month. His life was placid and he had a salary. In time, he travelled home to live beside the River Wei, returning to the capital just once or twice a month. Yet the palace library was proof that paper and ink lay at the heart of the Chinese empire. Anyone seeking senior officialdom had to be skilled at organizing and maintaining paper records. But the classics and calligraphy were just as important in securing paper's ascendancy, and their popularity was spreading to new segments of society.

Such was the ubiquity of learning in the capital that, even in the emperor's harem, women studied the classics, writing, law, poetry, mathematics and chess. They grew mulberry trees for paper and raised silkworms to supply palace silks. Imperial decrees written on yellow paper were sent around the empire. Army officers provided a stand covered with a purple cloth, a magistrate rested the official decrees on it and two legal mandarins read out each decree alternately.

Libraries spread quickly. The palace library held more than 200,000 volumes, representing an array of beliefs. By Bai's time, there were even translations of many Christian works, including the *Sutra of Jesus the Messiah*, which argued Christianity was no threat to Chinese culture but that Jesus was a saviour for all the nations, spurring Emperor Taizong to build a Nestorian church and monastery in the capital. Even in the furthest corners of the empire to the north-west, local government drafted its own paper documents. (In the Sui dynasty, during the sixth century, the civil service provided one emperor with 300,000 copies of an edict denouncing his rival.)

In 806 Bai Juyi had to study for another imperial examination, as a new emperor had come to the throne. As in previous exams, he prepared by sifting through the classics and commentaries on them as well as dynastic histories. But he studied for his exams in the library of Huayang, a Buddhist monastery – the flowering of Buddhism under the Tang saw new libraries spring up in monasteries and temples around the country, proof of the growing role played by paper in religious life as well as of the impact that written scholarship was enjoying on religious culture and priorities. Bai studied alongside his greatest friend, Yuan Chen. Each man had to explain the dynasty's decline and what should be done. Bai argued for low taxes before drifting into generalities about how a moral government should persuade the rebels to put down their weapons.

Yuan, however, proposed a new civil service exam system as the solution: first an exam on Tang laws and on the classics, then a second exam with answers written as prose-poems, as poems – and always in a more stylized format – were used for evaluating candidates. In the first exam, markers would give rote memorization a low priority in their marking scheme. In the second, markers would remember that this only showed literary ability, not policy knowledge or mandarin ability. There would be no further placement exam to allocate people to positions. In short, Yuan argued for a meritocracy not weighed down by classical thought and systems. Yuan came out at the top of all examinees, but his proposals never moved beyond answers on a page. Bai took second place.

Both had studied in a Buddhist library and read works by Confucians, Legalists, Buddhists and Daoists. Yet, as Bai grew older, his Confucianism began to dissolve and he turned increasingly to Buddhism and its sutras.

> There was a priest who brushed out sutras,
> His body pure, his purpose clear,
>
> . . .
>
> When the sutra was finished, he was named 'Sacred Monk' . . .
> On a white screen is Mr Chu's calligraphy,
> It's black as bold as the day it dried.

By the time of Bai, China's Buddhist clergy had become an elite of its own, second only to the state's secular hierarchy. This second elite had bred textual specialists, improved its own literacy and had passed new levels of literacy on to much of the Buddhist laity too. It was exempt from tax, pulled people from family commitments, undermined class distinctions and allowed women to take part in its ceremonies. It even diluted the country's Sinocentrism and yet the government during the first half of the Tang dynasty was eager to incorporate it into Chinese culture.

In the early Tang period, Buddhist monasteries had the largest libraries in China beyond the major cities. Their clergy was one of the most educated and literate groups in society and there may have been as many as a million Buddhist clergy in the mid-Tang. Nor were they merely holy men and women marooned in their monasteries; instead, many were deeply engaged in society – and in the marketplace. In fact, Buddhist institutions often were wealthy, thanks in part to slavery (during its persecution of Buddhism in 845, the government would confiscate 150,000 Buddhist slaves; many were sold, enlisted into the army or taken in by officials to work as their own slaves),[46] pawn-shops, tax breaks, land grants and their commercial lending businesses.

In 705 the emperor introduced exams for clergy. They had to answer questions and memorize passages from several texts, among them the Lotus Sutra. The 758 rebellion spurred the emperor to build a protective layer of Buddhist temples on China's five holy mountains and to appoint trainee monks. Each monk had to learn at least 500 pages of Buddhist texts – some sources suggest the number was closer to 700. Later in the eighth century, clerical exams were divided into disciplines and expanded to include hermeneutics as well as memorization. Monks also studied Confucian texts in their monasteries from no later than the fourth century. Monastic teachers taught calligraphy to their students too.

Even junior monks had to read and memorize a number of texts, and the youngest entrants, the six-year-old 'crow chasers', must have been taught to read. The Mahayana sutras repeatedly encourage people to read, recite, study and copy the scriptures. At the end of the seventh century, one temple in north-west China kept fifty-five copyists in work. The capital itself was surely home to several thousand copyists, working in its many monasteries.

The upper classes (and poorer civil servant families like Bai Juyi's own) received a Confucian education, but Buddhists probably focused on providing education in the villages. Some young students stayed in Buddhist monasteries to learn to read and write although many monasteries also commissioned murals to teach Buddhist beliefs to the unlettered masses.

A steady stream of sutras from India kept the shelves well stocked. Monasteries were producers of texts too, their monks doubling as scribes and often receiving government sponsorship to produce large collections of Buddhist texts. Aside from sutras, they also compiled histories of their orders and even copied many secular texts. Some of these they held simply to argue against: the court often sponsored lectures on the three teachings (Confucianism, Buddhism and Daoism) on the emperor's birthday and debates were common at such events. In the early seventh century, almost a million Buddhist scrolls were copied on imperial orders in the course of a decade-and-a-half.

Some, like Bai in 806, came to browse and to study a few of the thousands of scrolls stacked on the shelves of these monastery libraries as they prepared for their civil service exams. Chang'an alone had ninety-one Buddhist monasteries in the eighth century, almost a third of them for nuns. Even in seventh- and eighth-century Japan, monastic libraries held a number of Confucian works as well as thousands of Chinese Buddhist sutras and commentaries.

> I wander along
> Trailing my green bamboo stick
> And whistling the Daoist Yellow Court Sutra.

But Buddhism's greatest gift to literacy in China was neither libraries nor schools: it was reproduction. Monasteries had trouble producing Buddhist texts quickly enough, especially given the complexities of the Chinese script. In the cities shopkeepers would copy sutras too, selling them, alongside statues, to lay Buddhists. Buddhism, unlike Confucianism, did not set great store by fine calligraphy; its focus was on copying the words alone.

Bai's knowledge of the Confucian classics, his love for Buddhism, his service in imperial government and libraries and his poetry set him at the heart of China's paper culture in its golden age. His relationship with paper was not only one of the era's most productive, but one of its most enlightening and arresting. Bai was neither an emperor nor an official historian, yet it was his mix of inside knowledge and independent reflection that allowed the paper page to make his name live on down the centuries. Very few lives outside the imperial family were recorded in such detail, and almost none with such candour and emotional intelligence. In this sense, Bai stands as an exceptionally early example of an author using paper to describe his inner life and sometimes the details of his daily experience. Other available surfaces were too expensive, too rigid in format and too tied to old subjects and forms to promote a blossoming of such new forms of writing.

Yet it was Bai's literary loves that made him unlikely to appreciate paper's greatest partner technology, despite its appearance during his own lifetime. And even if Bai did not see printing as an important part of his own life, it was certainly crucial to the fortunes of paper before his death.

For at least two millennia intaglio seals (on which the image or character is cut out of the stamp) had been a mark of authority and identity in China. Emperor Shi (the 'First Emperor') had owned the first imperial seal, carved in jade, and Daoists had carved charms on wooden blocks to leave impressions in sand or clay, in order to expel evil spirits. By the time of Bai, seals were no longer intaglio (intaglio seals produce reversed images, like photograph negatives); instead, they were carved in relief such that the stamped image appeared on the page just as if it had been printed with a wooden printing block.

11 The calligrapher Yang Xin's personal seal, which appears on all his works. The culture of seals and of ink rubbing were both influential in the development of printing.

Bai himself probably used a stamp of his own on his poetry, a mark of his moral authority on the page. Daoists and Buddhists alike, and Bai knew plenty of each, also used seals on their writings. Bai would have seen plenty of rubbings too. Throughout Chinese history, hundreds of thousands of characters have been carved in stone and then – sometimes almost immediately, sometimes not for centuries – ink rubbings have been taken of the stone slabs on paper, creating a negative image of the text.

Reproduction has a rich history in China, from Shang dynasty bronzes to the script Emperor Shi standardized, to off-the-peg Ming porcelain to the tick-tock production of today's Chinese factories.[47] But mass reproduction was not part of Bai's own way of life, even if he saw it on his doorstep. Bai was a poet and for him, therefore, writing was brushed – a physical, idiosyncratic skill that expressed the character and emotions of its master. It took instead an ambitious empress to see where the future of mechanized copying might best be employed.

The first print run might have been a Buddhist project – the release of all Buddhist slaves in the ninth century had made hand-copying far less viable as a way of meeting demand for texts. But this project was not aimed at readers, nor would its deification of the empress have much impressed Bai Juyi. And yet the two lands where Bai won such large readerships in his own lifetime were (together with Korea) printing's heartlands: China and Japan.

In China, Empress Wu sent Buddhist texts to her favourite holy mountains, arming them with tales of the healing powers they carried. She ordered embroidered copies of 5,000 scrolls produced, more than 20,000 scrolls written as karma for her late parents, and several million amulets. In her palace, the women of the powerful Liu family were dyeing delicate patterns onto textiles in the early eighth century and eventually the empress was given a print-dyed garment of her own.

Wu used sutras not as texts to be read but as charms to propel her image, and the project of mass production on which she embarked may have sparked the transfer of printing from fabrics to the paper page. In fact, the oldest surviving printed paper in the world was not even designed primarily to be read: a Chinese Daoist invocation or charm, probably printed in China, but on Korean paper, in 751. Its purpose was as a spiritual talisman as much as a medium for words.

The earliest known print run was a political stunt too. In 770, the Japanese Empress Shotoku supposedly (and perhaps actually) had a 'million copies' of a Buddhist word-charm printed on miniature sheets of white hemp paper. Each was placed in a tiny stupa and sent to temples across the islands to bless her reign. Bai the civil servant would have been unimpressed; he might even have composed a poem to undermine such vanity. Yet Bai was a Buddhist too, latterly, and the world's first printed book, which appeared just twenty-two years after his death, might have pleased him. After all, this one was actually designed to be read.

In 868 the Diamond Sutra was carved onto wooden blocks in Chinese characters. The blocks' faces were covered with ink and pressed onto a series of sheets. The first sheet of paper was imprinted with a woodcut illustration, showing Sakyamuni Buddha himself surrounded by a cartoon crowd of followers. The printer placed this sheet at the front of the sutra scroll, pasting its left margin to the first sheet of text so that the scroll was read from right to left. After the sheets of paper had been made, they were dyed with berberine, a dye which turned them yellowish and kept insects at bay. They were then printed onto with durable, carbon-based ink before being glued together to form the single scroll that survives today. Unrolled, this copy of the Diamond Sutra stretches more than sixteen feet. It was, of course, a crucial departure from the handwritten scroll, whose identity could not be separated entirely from the scribe who had brushed the characters out on its surface. Nevertheless, printing grew out of those same obsessions that had propelled the written text to become so widespread in Tang society: a love for paginated knowledge and reflection, whether in government papers, religious instruction, philosophy or poetry.

Buddhism brought printing to East Asia, but it failed to exploit its potential. From the mid-ninth century, the religion declined in China, falling victim to popular waves of Chinese nationalism and xenophobia, as well as to state-led persecution, the destruction of monasteries and an end to tax breaks. Thus it fell to the civil service bureaucracy to adopt printing for itself and the Imperial Records Office did indeed begin to expand, producing ever more papers of state. Mandarins in the Tang period already needed to excel in managing paper records but printing delivered new productive capacities, especially for the Song dynasty, which succeeded the Tang in the early tenth century.

Even before printing enjoyed its full impact, the volumes of papers and expectations of written reports were central to the life of a civil servant. In 808 Bai Juyi told a state archivist of his own monthly paper load – 200 sheets – in a 1,000-character poem. Bai said that he had written it drunkenly and with a 'racing brush':

> In time receiving two hundred sheets of protest paper monthly,
> Ashamed at my salary of 300,000.

Two years earlier Bai had worked as the Director of Administration in a provincial town in the centre of northern China. In 807 he became a proof-reader in the Hall of Assembled Worthies, where he helped to compile excerpts from the classics as well as commentaries on them. It was a minor role but the emperor began to notice Bai. The same year a eunuch came to his house to tell Bai he had been appointed a Hanlin scholar, one of the most prestigious positions available at court.

As a Hanlin scholar, he worked as a private secretary to the emperor, drafting and preparing documents for him to read, approve and then stamp with his vermilion seal. More than 200 of these papers survive. Bai found close friends among his peers, all of whom were chosen for their writing and drafting skills. As Bai entered the Hanlin Academy, its membership was undergoing a cleavage between the new wave who wanted to stem the power of the eunuchs, and those who wanted no such fight. Bai was one of the reformists, eager to restore power into trained hands.

In 808 he was appointed one of six Imperial Censors to the Left, who formed part of the Imperial Chancellery. The position was not especially senior in terms of rank (he was now in the third class of the eighth grade) but the job itself was privileged. It allowed Bai to write 'memorials' (official records) for the emperor to read but also to approach the emperor in order to criticize his deeds, words or proclamations and to suggest reforms to government practice. It was the closest he would ever come to the world of policymaking. Bai believed that his poetry was the reason the emperor had chosen him over several of his peers. He also believed in the possibility of the emperor improving the lives of the poor, although sometimes the emperor's actions came far too late.

> The emperor published his wise decree on white hemp paper,
> For the district to be free of taxes for a year . . .
> But for every ten families, nine had already paid . . .

Bai was only meant to criticize the emperor's own actions, but he had seen destitute farmers in northern China as well as corruption at the palace among eunuchs, mandarins and the emperor's own wives. In his new position he had legal immunity and did not have to produce any evidence to back up his reports. In one of his poems, malign wives and silver-tongued ministers are like pretty wisteria that twists itself round a tree before strangling it. In another, good officials are like lotuses in a pond that is so filthy they can't even produce their fragrance. Bai had been a politically active reformer for years already and he hoped to point the emperor to the true problems beyond the palace gates. In 809 he explained his role in verse. Brushes from Xuancheng in central China were, Bai wrote, expertly assembled, their creators choosing 'one strand from a thousand' from the hares they caught for the purpose. A warning follows:

> Designed for court, these brushes must not
> Be carelessly employed by emperor or advisers;
> They are best used by censors and historians
> To uproot evil men, or to capture
> The happenings of the time;
> Censors and historians alike must see
> These brushes as invaluable, employing them
> Not to censor trivial faults, or to recount
> The cheap wording of a minor command.

The same year, Bai advised the emperor to revive the practice of collecting folk songs. A millennium earlier, in the Zhou dynasty, the Folk Song Collector travelled through China to write down popular folk songs and in this way popular feeling would, it was hoped, reach the throne.

> . . . those songs still sung
> Beneath the altar all praise
> The emperor; poems and prose
> Extol him, whereas good advice
> And disapproval is, these days,
> Absent; mandarins employed to deliver
> Counsel keep their mouths shut
> Avoiding word or deed;
> The Great Drum, for petitioners
> To beat, hangs beyond reach...
> I pray the emperor hears me now.
> If he hopes to appreciate
> What the people think and feel,
> He should ask those equipped to criticize
> To cage their thoughts in poems.

Bai did just that, and warned the emperor that the manipulation practised by the palace eunuchs, the wilful disloyalty of regional governors and the rural poverty that they caused would lead to the downfall of the Tang dynasty. He argued that the value of a copper coin was greater when melted down than when used as money. He advised that all copper containers should be confiscated to control the price of money, since fluctuations in purchasing power were harming the poor, but his proposal was ignored. As farm produce dropped in price, Bai complained to one friend that agricultural taxes did not, such that the rural poor suffered. He even wrote a poem condemning the annual tribute of dwarfs sent to the court by officials in Hunan, far to the south. (Tributes arrived in the capital from various vassal states as part of a patronage system and often included the products for which an area was renowned. Dwarfs from the south were just one of these tributes – human curios to amuse the emperor.)

These years were Bai's most political and patriotic, and he lived out his views and moods on paper throughout. He was thirty-seven years old, residing in a quiet area of the capital with the woman he had married a year earlier. In one poem, he described a typical morning. He rose before dawn to greet the emperor, walked to the Silver Gate Terrace along the banks of the Serpentine River, travelled three miles north on horseback in a bitter wind, and then waited for the water clock outside the city gate to chime as his moustache gathered ice. At the end of the poem, he intimated his own future when he imagined his friend Chen still sleeping under furs until the sun was far overhead.

In addition to the emperor's trust, Bai also received the support of the man appointed as prime minister in 808. Bai's friend Yuan Chen was given a senior appointment by the new prime minister too. Throughout 809 Bai's poetry was turned to politics and the few personal poems he penned were usually about or for his friend Yuan. But one day, as he crossed the city, Bai ran into him unexpectedly. Yuan had just been exiled, the result of accusing a recently deceased but popular official of tax corruption and illegal property confiscations, stirring resentment against him among other officials. So it was that Bai rode south with his friend for some time before he had to turn back. But he used the official messenger service (which spread across the empire over the course of the Tang dynasty, utilizing both land and water routes) to deliver a scroll of twenty new poems to his departing friend. Bai's letters of protest to the throne failed to reverse Yuan's exile, however.

Bai quickly found enemies among the eunuchs and the older bureaucrats, yet he continued to take risks with his advice. When a matter was too sensitive even for him to raise directly, he would write a poem about it and distribute it anonymously, in the hope it would grow popular enough across the capital to reach the emperor's ears. (His most famous English-language translator, Arthur Waley, called this his 'letter to *The Times*'.) In one of these poems, Bai pictured men from the Board of Punishments and Justice Ministry enjoying a party together one evening, unconcerned that inmates at one of their prisons were freezing to death.

In some ways, Bai was just one of many concerned scholar-officials looking for ways to criticize imperial policy without losing his job or his life. Through imperial China's history, poetry became the most popular way to do this, and a reference to a flower in bloom, an emperor from ancient times or changes in the weather could all be used as political satire. The educated class was eager to read Bai's poems, after all. But it was a high-risk game.

Bai's 3,000 surviving poems stand as the work of an Everyman, and of someone eager to help and appeal to the peasant farmer. According to a later poet, Bai would read all his poems to his washerwoman before publication to ensure they were properly understood. But when it came to officialdom, his illustrations were often damning, as in one poem he wrote in 809 about the trouble officials took to find the best carpets to please the emperor.

> A few feet of this carpet
> Needs a thousand *taels* of silk;
> Floors do not feel the cold.
> Best not to rob the clothes
> Off the people's backs
> Simply to lay them on the floor.

In 810 Bai asked to be moved to a better-paid position that would allow him to look after his ill mother. In the autumn he left on sick leave and stayed away until the end of the winter. As he spent time in the hills, he began to wonder whether nature and rural living were his vocation. Moreover, Bai's mother drowned in a well the following year, so he gave up work to carry out the Confucian three-year mourning, leaving the capital in 811. Bai's letters at this time suggest that his political hopes had started to fade. He even began to speak of his career in the past tense.

In 811 Bai's child, a girl called Golden Bells, died at the age of three while he and his family were living close to a western tributary of the Yellow River. Bai's Confucianized view of women seems to be on the point of unravelling as he writes of his daughter's death:

> A daughter can snare your heart;
> And all the more when you have no sons.
> Her clothes still hang on the pegs,
> Her useless medicine lies by her bed.
> We saw you off down the village street
> And watched them pile the earth on your grave.
> Do not say you're just a mile away –
> Between us now lies eternity.

Bai returned to the capital in 814 as adviser to the crown prince, but the following year he was banished for writing two poems that were seen to lack filial piety, a foundational Confucian virtue. More importantly, the political poems in his *New Music Bureau* collection had criticized too many senior officials, thus offering negative commentary on the emperor too. These poems complained of the imperial harem of fifty years before in order to make a comment on the size of the present one. Bai disliked the way women were confined to the harem for life without a future, a family or a husband. He had suggested that tax levels and harem numbers be reduced and the emperor agreed to his request – the rains that followed were thought to prove Bai's Confucian rectitude.

In 814 Bai had written to Yuan Chen that their hearts were fused in their debates about writing and that his sleeves were filled with writings. The following year Bai sent his old friend a thirty-page letter about poetry, proof again of the depth of self-expression that paper had handed such men. (Such letters were a remarkable feature of Tang life for officials and literati, following the development of the dynasty's postal service. Moreover, the Tang postal system was crucial for the spread of poetry, enabling the emergence of poetry groups and allowing contemporary poets more immediate geographical reach.[48])

In his letter to his friend, Bai said he had come across the twenty-six scrolls of Yuan's writings and, as he read their contents, it had felt as though Yuan was with him. He claimed the Confucian classics as the sun, moon and stars of human life, headed up by the ancient *Book of Songs* which crystallized the feelings of the people in verse. But a bad crop of poets had since written about nature for nature's sake and ignored poetry's role in politics; even Li Bai and Du Fu, two of the greatest, wrote only occasionally about politics.

> It was then that I came to this conclusion: the duty of literature is to serve the writer's generation; that of poetry to influence public affairs.

In the letter Bai recounted the many meetings on his travels with people from the capital who knew his poems: singing girls, schoolmasters, Buddhist priests, government officials. He found his poems written on the walls of taverns and the sides of ships, and recited by mandarins, monks, old widows and young women. Bai wrote that his fame had balanced out his suffering. His new focus was to collate his poems, splitting them between didactic, reflective, sorrowful and miscellaneous. (He complained that people only remembered the fourth kind.) He went on to remind Yuan (who he calls Weizhi) that their whole friendship had been carried on in verse (he might have added 'on paper') before ending the letter intimately, referring to himself by his more familiar name, Letian (meaning something like 'happy-go-lucky').

> Picture me this evening, brush in hand, paper laid out before me, as I sit by the lamp. Silence surrounds me. As thoughts materialize, I set them down, not seeking to order them. This long and garbled letter is the outcome . . . But Weizhi! You know my heart well enough! Letian bows twice.

In 817 Bai built a small cottage below Incense Burner Peak, close to the Forest Temples at one of the five holy mountains of China, and limited himself to four wooden couches and two screens, as well as a small collection of Confucian, Daoist and Buddhist books. He began to settle into quietude. Here he received several hundred poems written by Yuan Chen and decided to use them as his furniture.

> Remembering you, I choose your poems for a screen,
> I write and study them without distraction,
> When the screen is ready, everyone will copy it,
> And the price of paer in Nanzhong will soar.

> In days past I sealed a letter I'd written you one night
> Behind Golden Bells Pavilion as night was turning to day.
> This night I seal another, but in a hut, and on Lu Mountain.

Just a couple of years later, however, Bai was recalled out of exile and back to the capital, where he filled one position after another. The following year his friend Yuan, now also back from exile, was made head of the Hanlin Academy (although the appointment was short-lived), often referred to as 'the forest of brushes'. Bai was now transferred to become governor of Hangzhou, near the Yangzi Delta. When his friend Yuan Chen visited, their friendship was already so famous (thanks largely to Bai's poems) that crowds gathered to see them stroll out together. (Bai and Yuan's poems would be among the first to be mechanically printed and produced in large numbers.) As governor of Hangzhou, Bai built a flood dyke to help the farmers. He said when he left that it had been his only real achievement there, yet residents lined the streets to mourn his departure in 824. When he arrived at Luoyang, where he had a new position working for the crown prince, Bai wrote that he was growing old. He recounted how he had written more than a thousand poems on nature and on customs everywhere, but that he remained as obsessed with poetry as ever.

Bai saw exile, age and enemies cull his political ambitions. He had spent a decade studying the classics on paper so that he could enter the civil service. He had passed another two decades shifting paper inside it and writing poems on paper against court corruption and about the suffering of China's farmers. Bai personified his age, flush as it was with art, learning and wealth, but he also saw that it was an age owned by aristocrats and officials. Corruption was cracking the pillars of state, and the poor were not invited to share in the Tang's successes. Bai had finally tired of official life. Exile, among the greatest themes of classical Chinese poetry, had begun to convert him to nature and to idleness, although idleness meant poems and correspondence, especially with his beloved Yuan.

> Red headers, white paper, two or three volumes;
> Half are your poems, half your calligraphy;
> A year of illness just passed,
> Now opening them
> I find bookworms have eaten them.

Yuan was a fellow graduate, official and poet, probably held back in his career by his candid poems as much as was Bai. But Bai often turned to Yuan's scrolls of poetry when he was alone and ostracized, as he was in 815, sitting in a boat on the way to the city of his exile, Jiangzhou.

> Picking up your poems to read by candlelight
> I finish them, the candle now short, the night as deep as ever.
> My eyes sore, I put out the light and sit in the dark;
> Winds churn the waves, water strikes the ships.

Their separation was unavoidable. Chinese officials were sent to cities all over the country once they had graduated and it was rare for them to see each other again. Nevertheless, these student friendships were more highly valued than any others, and for Bai and Yuan, one another's paper poems became a substitute for true companionship.

> You brush out my poems on the temple walls,
> And I fill screens just here with yours.
> Dear friend, who knows when we'll meet again,
> Two floating fronds, far out at sea.

The range of experiences (censure, loss, success, love) and roles (civil servant, poet, exile, friend, mourner) that Bai lived out on paper is remarkable, even to modern readers. The Chinese characters had become the means to significance and Bai learnt them from his early youth. The Confucian classics were the only way into the civil service and Bai studied them on paper scrolls in detail. But much of his study was in a Buddhist monastery and Buddhism had begun to spread literacy more widely and to develop printing for the page. Bai worked in imperial libraries, collating the classics, collecting commentaries and writing summaries. He also recognized the poverty and suffering of farmers as well as the corruption of the court and its eunuchs and he sought to redress them with written complaints, on paper, to the emperor. He poured out his heart in poems and letters to close friends. Bai's thousands of poems offer his readers a life expressed on paper.

> While I live, I will not grow rich or win honour,
> But when I die, I know my books will live.
> You must forgive me this petty boasting –
> Today I added Volume Fifteen to my works.

In old age, Bai learnt to love idleness. His servant would bring his bowl and comb to his bed, set his chair in the sunlight and place warm wine and poetry books on the table. He would go for walks in the mountains. When his left leg was paralysed by a stroke in 839, he revised his poetry collection. It now contained almost 3,500 poems and prose pieces. He sent copies to five different monasteries, apologizing for profaning their libraries with irreligious works.

Bai outlived most of his friends; in 831, his dearest friend, Yuan Chen, died after suffering an illness that had first struck him only the day before. Nine years later, Yuan's memory was unexpectedly resurrected.

> His poems are long since buried
> Deep in cupboards and cases.
> But just recently, someone was singing
> And I heard a line –
> Before I could make out the words,
> A pain had jolted through my heart.

Although Bai loved the idleness and escape of retirement, even at seventy-three he was still concerned to have a public waterway cleared since it was a danger to river-traffic. Bai called writing poems his affliction, his obsession and his instinct; they were, perhaps, his way of remaking the world as it should have been. In 837 Bai's daughter had a child and his last surviving poem, written in 846, has Bai in bed, still talking of his beloved poetry.

> My bed is moved up the plain screen;
> My stove placed in front of the blue curtains.
> Grandchildren read to me from a scroll
> And a servant prepares hot water.
> With racing brush I answer the poems of friends,
> And search through my pockets for medicine-money.
> When such trivial matters are dealt with,
> I lean back on my pillows, with my face to the South.

8

The Handover

For knowledge, go even unto China.
(traditionally attributed to Muhammad)

Bai and his contemporaries were part of the great flowering of China's paper culture across government, scholarship, poetry, religion and the marketplace. Their achievement marked the apogee of paper culture in China; in centuries to come, it was the Tang dynasty to which China would look back as its imperial golden age. Poets in particular found it difficult to emerge from under the shadow of the Tang greats, of whom the best-loved were Bai Juyi, Li Bai and Du Fu. There were other paper disciplines in China that would find their greatest form under later dynasties, notably landscape painting, which particularly flourished from around the end of the Tang dynasty (which came in 907) to the early twelfth century. But the Tang era would resonate with future generations for its poetry and calligraphy.

Paper, then, had established most of its pre-modern Chinese identity by the middle of the Song dynasty (960–1279). As a vehicle for calligraphy, poetry, painting, religious writings (and proselytization) and government bureaucracy, it had been thoroughly tested and found to have no serious rivals. It had other functions too, of course, from toilet roll to kites to window panes, but its major uses in pre-modern China were written and they were now largely set. Significant change would only come in the modern era; at first through some of the revolutionary periodicals and newspapers published late in the Qing dynasty (1644–1911), which helped to precipitate the fall of China's two-millennia-old imperial system and to forge an urban readership, introducing that readership to political debate and informing its members of political developments in foreign nations. Supremely, however, paper's new role found expression in the May Fourth Movement of 1919, when students marched in protest against the government of the newly formed Republic of China because it had signed the Treaty of Versailles, which handed control of the Shandong peninsula (south of Beijing) to Japan. But if the loss of Shandong was the catalyst for protest, the demonstrators had other aims too: to modernize China and usher in a new era of equality and opportunity. They aimed to rouse 'the masses all over the country' and 3,000 of them marched in

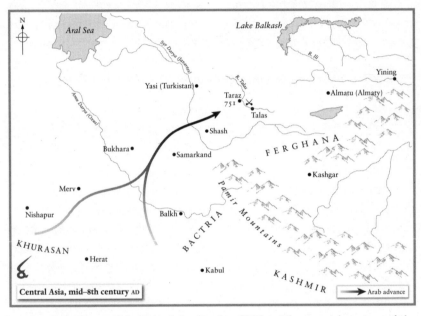

12 Central Asia at the time of the Battle of Talas. The exact location of the battle is disputed, but was close to the present-day towns of Taraz and Talas.

a procession, handing out printed paper broadsheets calling on all Chinese to protest. Their protest spread through the country. In Shanghai, 60,000 workers set down their tools and left their factories.

The more educated protestors had read or even watched modernist Western plays – Ibsen-mania was exceptionally strong – and dreamt of emancipation from the ties of colonialism, class and sex. In *May Fourth* journals and broadsheets (the 1919 protests began on 4 May, a date which subsequently became shorthand for the whole movement), articles and essays were written in the vernacular. The protestors' audience was the Chinese nation and one of their aims was the widespread adoption of the spoken language on the page. The movement had the support of intellectuals like Hu Shi (later a nominee for the Nobel Prize for Literature), who complained that a dead language was incapable of producing a living literature.

Another man who argued for change was Liu Bannong. Liu was one of the scholars whose influence would one day spur the introduction of simplified characters in China, a crucial step in making the Chinese script accessible further down the social scale. In 1918 Liu's translation of the poem 'I Walk in the Snow' was published in the influential Chinese magazine *New Youth* (Xin Qingnian). In the Preface to his translation, Liu said that he had tried different translation techniques, but that the Buddhist approach had helped him most:

> When I read this poem in *Vanity Fair* two years ago, I tried first to translate it into traditional poetic forms . . . but failed due to the strictness of their styles. Then I used the method of translating Buddhist scripture and conveyed the sense directly . . .

Liu and the reformist students won their fight to modernize the language on the page. By the end of the 1920s, the phrases and constructions of everyday Chinese were officially accepted on paper. Books, legal and government papers, and newspapers now appeared largely in plain speech as Everyman trumped the Confucian. It was a trend that had begun, as Liu understood, in the first centuries AD, with Buddhist translations that used ordinary Chinese phrases and sentences to reach a broader audience. As then, words and names were transferred directly from other tongues into written Chinese. When twentieth-century revolutionaries composed Chinese translations of Western novels, academic works and poems, many turned to the old Buddhist translation methods. They were, in this sense, the heirs not just of the European modernist revolution, but of a far older, Chinese textual revolution, one which paper Buddhism had sparked.

Printed European books and Western-style newspapers started to be published and read in China in growing numbers in the nineteenth century. And with this wave of Western influence, the wheel had come full circle: China, which had first delivered paper to the world, was now reshaped by paper imports from Europe and America.

In a sense, then, the rest of this book chronicles how such a reversal was able to come about. Tang China was, admittedly, geographically larger than ever and enjoyed unprecedented influence abroad, but even the Tang's borders remained thousands of miles from Europe and the two continents had no direct interaction. And yet it was during the golden age of the Tang dynasty that papermaking, apparently without the knowledge of the emperor's court, finally broke loose from East Asia. The handover began, one of the greatest Muslim historians tells us, with a battle at China's outer reaches.

The River Talas flows from the Kyrgyz Mountains, 200 miles west of China's modern borders, to the Muyunkum Desert in Kazakhstan, and forms a stitch across the centre of what used to be called Turkestan, the lands of the Turks (or Turkic peoples). From the eastern shores of the Caspian Sea, Turkestan stretches eastwards, ending some 200 miles short of the west end of the Great Wall of China. Over the centuries, the region's deserts, mountain ranges and trade routes have been periodically sliced up into different states and, more recently, its mines, oilfields and gas wells parcelled out between them.

In the seventh century, however, the Turkic peoples were united as the sole regional power, albeit with the armies of Islam pressing in from the South. In 734, when the Turkic khan was poisoned by one of his own men, it spelt the formal dissolution of a large kingdom that had once been on the verge of empire. Even five centuries earlier Turkestan had already become the world's crossroads, its network of highways serving China, India, Persia and the Roman empire. Silks went west while gold and wool travelled east. Precious stones, diseases, satins, religions, horses, languages, slaves, scripts and ideas furrowed across Eurasia and the Turks became the middlemen of a set of global trade routes that linked Chang'an to Rome: the Silk Road.

Yet it was a small-scale military encounter in no-man's-land which would be hailed as the crack in the dam that had confined papermaking to East Asia for centuries. While some Chinese papers were imported into Central Asia and Sassanid Persia as luxury items in the seventh century, paper was only commonly made in Central Asia after the Battle of Talas, which took place in 751 on the banks of the Talas river. The sale of a few bundles of luxury paper was far less significant for the recipient culture than the transfer of the practice of paper-making itself. A society might be able to enjoy using expensive, imported Chinese papers on a limited basis, but paper could only become integral to a culture if the manufacturing know-how was first imported and then widely adopted.

Unlike the goods they sold, few individual merchants travelled very far along the Silk Road – instead, goods were sold on from one mer-chant to another. Given these constraints, the transfer of goods was far easier to facilitate than the transfer of ideas and technological processes (unless, of course, those ideas and processes were written down). Moreover, a commercial pass-the-parcel is an ill-fitting expla-nation for what was a rapid transfer of papermaking know-how from Greater China to Central Asia and Persia. The earliest surviving explanation for the transfer, written in the eleventh century, certainly viewed the shift as rapid – and decisive.

For a small battle, Talas could not have been fought between two greater powers: the Muslim Caliphate and Tang dynasty China. In 750 a new dynasty claimed the Muslim empire and founded its capi-tal at Kufa, a little south of present-day Baghdad – the Abbasid Caliphate. From Kufa, it ruled Mesopotamia, Syria, parts of the for-mer Byzantine empire, the entire Arabian Peninsula, the full breadth of North Africa, the Iberian Peninsula, the former Persian empire and much of Central Asia. As its rule spread out to three sides of the Mediterranean, it spurred one of history's greatest human migrations (as Arabs settled in conquered territories, like Syria and Persia, across the *Dar al-Islam* – the lands of Islam) and it produced entirely new heights of art and advances in the sciences. But the Caliphate had an equally impressive adversary to the east.

In the mid-seventh century the population of the world was less than 500 million, and China alone accounted for 50 million people. China controlled the eastern Silk Road, enjoyed good relations with its major nomadic neighbours, kept a large army, tapped into a tax register that listed more than 800,000 families (although, like China, this would splinter as the Tang dynasty progressed), and conducted trade by sea and land across East, South and South-East Asia. In China itself, a 1,000-mile canal held the country's north and south together, bringing grain, cloth and money to the capital in a steady stream. China sat at the centre of East Asia, an image of unity, power, wealth and artistic brilliance.

In the first half of the eighth century, both China and the Islamic Caliphate made bids to control Turkestan. China used both frontier garrisons and protectorates in Central Asia, and by the 750s China was ruler of the Tarim and Issyk-Kul valleys that straddle its modern western border. It was commander of the Pamir Valley, protector of Bactria (northern Afghanistan and the regions that border it) and ruler of Kabul and Kashmir.

The armies of Islam, on the other hand, had taken much of Transoxiana and had begun to pick off some of the Chinese garrison towns too, moving into the east of Turkestan. Perhaps more importantly, much of Turkestan had been converted to Islam through a mix of force, job discrimination, tax discrimination, the choice of death-or-exile for polytheists and atheists, sporadic persecutions, the destruction of old temples and idols and an impressive preaching programme. The Caliphate's rule was increasingly proving impossible to resist.

It was then that General Kao, the Korean-born commander of China's western army, made an important mistake. There are different stories of how he came to capture the king of Shash (now Tashkent), an important local ruler. What is clear is that he sent him back to the Chinese capital (despite having possibly offered him a hollow pardon in return for surrender), where the king was put to death.

As a result, the late king's son recruited the Arabs to help avenge his father's death. Abu Muslim, the Arab ruler of the Eastern Caliphate, sent his best general to fight the Chinese armies. At this point, the grieving son pushed the Arabs to aim not for Ferghana, China's major regional ally, but for the four cities of eastern Turkestan that gave China a vista on the whole region: Kuche, Kashgar, Khotan and Karashahr. But the Arabs had chosen to bypass Ferghana anyway. They knew that if they could defeat the Chinese further east then Ferghana would be cut off and would also fall into their hands. They had chosen Talas.

*

In 751 the Chinese and Arab armies met on the banks of the Talas. The river runs westwards for more than 250 miles (although less than 200 as the crow flies). It springs up in the Kyrgyz Mountains and, before it reaches the border with Kazakhstan, nearby peaks spike to over 8,200 feet on its north side and to over 6,500 feet on its south bank. Between these skyscrapers, the cracked course of the Talas Valley winds towards the desert. Yet there are patches where the valley broadens enough to accommodate thousands of men; it was on one of these plains that the armies gathered.

The Chinese had 30,000 men, according to their official histories. (In the Arab histories, they had 70,000, but exaggeration of numbers was not uncommon in early Arab histories.) Twenty thousand of the soldiers were Chinese and most of the rest were Turkic. The Chinese army included men from Ferghana too, but all had travelled, on foot or by horse, for close to 200 miles. Many would have needed rest. The Arab army that opposed them numbered 40,000 men. It was close to an even match.

The Arab soldiers and cavalry were fresh and were excellent horsemen, and they were reinforced by Turks from Shash and the surrounding area. The Chinese infantry and cavalry had endurance and professionalism on their side. Their Turkic allies were the best mounted archers on the battlefield. Like the Arabs, China's army fought with bows, spears, swords and probably emblazoned shields. Chinese soldiers often wore boiled leather armour too.

General Kao, like his opposite number Zayid ibn Salih, was considered to be an excellent general. He had led stunning campaigns in 747 and 750, although the emperor may have been the puppeteer who ensured his success.

The battle lasted five days. As infantry soldiers alternated assaults with defensive spells, the cavalry roamed the fields, probing enemy weak points, releasing arrow showers and exploiting small victories. The meeting of Asia's two great empires looked likely to fizzle to a truce or even to abandonment. Instead, the Karluk Turks, hitherto Chinese allies, attacked the Chinese army from the rear. The Karluks were significant local players from the area close to Lake Balkhash (now part of Russia). Karluk soldiers within the Chinese army probably switched sides at this point too. (General Kao's betrayal of the king of Shash may have been what persuaded the Karluks to fight for the Arabs, as the Karluks had supported his rule.) Some 20,000 Chinese soldiers were killed and General Kao had to forge a path through his fellow fugitives to escape with a few hundred men.

The Arabs might have pushed on and exploited their victory, but they were weakened too. The residents of Bukhara had recently rebelled against their rule and there was concern that the Chinese populations of cities further east might cause problems. So the Arab armies returned to their Central Asian homes. The Battle of Talas is sometimes touted as the reason Islam became the religion of Central Asia. China would not make any significant territorial claims beyond modern Xinjiang for almost a millennium afterwards.

Yet in many ways nothing had changed. The Chinese had lost 20,000 men, but the four crucial cities of eastern Turkestan remained in their hands. Their failure to push into Central Asia in the centuries that followed had more to do with rebellions, divisions and invasions within the Chinese heartlands than with the defeat at Talas. As for the Arabs, they had simply preserved the wide buffer between themselves and China's western reach. In military terms, Talas was more a curious detail than a turning point.

But as the Arab soldiers returned to the cities of Central Asia, they took captives with them as booty, many of them to Samarkand, the city once described by Alexander the Great as the most beautiful on earth. Slaves brought Chinese knowledge and skills to their new homes and masters and, according to the eleventh-century Muslim historian al-Tha'labi, among these was a process that had remained cocooned in parts of East and South-East Asia for centuries. For as the slaves settled into the cities of Transoxiana, they began to teach the Arabs how to manufacture paper. Moreover, parts of the region would have been very receptive to the use of paper: the seventh-century Chinese Buddhist traveller Xuan Zang reported that boys in Samarkand were taught to read and write at the age of five.

There were papers in Central Asia slightly earlier than the battle (and even one collection, found in Gilgit, Pakistan, from the sixth century), but very few survive and they either represent only the hesitant beginnings of local production – or else were simply Chinese imports. The first Arabic word for paper, *kaghad*, comes from regional languages of Central Asia (Sogdian and Uighur). The flowering of paper culture in Central Asia after the Battle of Talas makes it hard to dismiss the theory out of hand. Coming from one of the greatest Islamic historians, it remains the best explanation we have and, even if not true in its details (as many have argued), the story's timing and geography make very good sense. Al-Tha'labi himself was in no doubt that the battle on the banks of the Talas, that lone armed meeting of sprawling empires whose capitals lay almost 4,500 miles apart, was the spark that sent papermaking on its trail across Asia – and so around the world.

9

Bibliophiles

I wish the Zindiks [Manichaeans] were not so intent upon spending much money in buying clean white paper and using shining black ink, and that they would not lay such great store on beautiful script, and in inciting their scribes to zeal; for truly, no paper that I have seen is comparable to their books and no beautiful script with that which is there.
The Islamic scholar al-Gahiz (d. 868) reports what his friend Ibrahim al-Sindhi once said to him.[49]

As papermaking stepped out beyond the lands of Greater China, it would need new sponsors. One of these would be a new religion obsessed with the artistic possibilities of the paper page. The other would be a people, the Turks, with a very different history, outlook and culture from the Han Chinese. And yet, to some Han Chinese in the early Tang, the Turks were something of an obsession.

In the 630s, China's crown prince, Li Chengqian, had put up a tent in the grounds of his palace in Chang'an and moved in. He had placed wolf-head banners outside the entrance, had chosen attendants who looked like the Turkic Uighur people of China's north-west and had begun to learn the Uighur language. He had ordered lambs to be roasted whole, had carved the flesh off roast sheep with his sword and had made his attendants braid their hair, wear sheepskins and work as shepherds in the palace gardens. He had re-enacted a Turkic funeral by lying on the ground and playing dead while his attendants rode around his body shouting. The heir to the most settled civilization on earth had remade himself as a Turkic khan living in a tent.

The Uighurs were China's other, Gypsies to Europe's settlers, and thus, to the early Tang dynasty Chinese, they became a national muse. In the markets of Chang'an, foreign clothes and curiosities were two a penny. Persian felt hats arrived from Central Asia, men's leopard-skin hats were sold alongside ladies' hairpins that trailed noisy trinkets. Women wore tight nomad collars and sleeves and, in the early eighth century, began to wear men's clothes too, like the nomads of China's north-east. Turkic musicians played their harps at the imperial court. In writing, Chinese had several scripts to choose from: fallen shallot, tiger claw, wind, moon or ripple. But many people were curious about the foreign scripts and books they encountered too, especially if those scripts and books were from the Turkic north-west.

The north-west of modern China is called Xinjiang, the new territory. The Uighurs ('wee-gers') moved there from the steppe further to the north in the year 840 and have remained ever since. When the armies of the Tang could not man their own defences, Uighurs were drafted in to help. They won special trading rights as a result, and many Persians and Central Asians in the capital adopted Uighur clothes and hairdos.

Many of the capital's taverns were run by foreigners (and loved by poets). They lay along the south-east wall of the city and were staffed with blonde women from Central Asia. The women had green eyes and white skin and they sang and danced at the taverns to keep their patrons buying ale and wine. Meanwhile, the Uighur men drank heavily and, over the years, proved themselves to be too boisterous for the urban Chinese.

The Uighurs were horsemen, drinkers, fighters and travellers, yet they had come to love books too, thanks in great part to the influence of Manichaeism, a religion which had arrived from the west. And in time they grew obsessive about writing and copying, calligraphy, paintings, papermaking, bookbinding and reading. This shift was in part thanks to their position as Asia's market middlemen, a benefit of their ninth-century move into north-west China. Since east–west trade routes crossed through this region, avoiding the mountain ranges to the south and the uncharted tribal areas to the north, the Uighurs were able to settle, accumulate wealth and enjoy the ideas and religions that washed into their towns from distant lands. Their celebration of the paper page for writing, and for much more besides, turned the book itself into something deeply sacred and desirable. Paper's standing rose as a result. As late as the 1910s, a Russian geologist living and working in Chinese Turkestan noticed that the local Uighurs placed all finished papers into small bins so that they could not be used for unclean purposes.

Even in the middle of the twentieth century, most of the region's Uighurs were illiterate. But many Uighur groups had converted to Islam (while a few regions remained Buddhist or Manichaean), and revered the written or printed word as a bearer of spiritual power. Indeed, for some of them, the Koran was not simply the written word of God but had magical powers too. Fortune-tellers used it for soothsaying, and others harnessed it, alongside a litany of other publications, to win new business, improve their health, tell fortunes, find wealth, and to encourage love in someone who was indifferent (known as *isitma*, heating) or its opposite (*sogutma*, cooling). Still others would jot a cabbalistic formula down on a scrap of paper, attach it to a thread and bury it in a cemetery to kill a rival. Paper became the means of selling, convincing, defending, cursing or blessing. It was not simply a surface to carry words but, with the right written formula, it was clothed with spiritual power.

The Uighurs were not Confucians and their own herding heritage had little in it to suggest a bookish future. But they became the conduit for paper culture to move westwards, and their great influence was neither a Chinese Confucian scholar nor a Central Asian Buddhist missionary, but an Iranian named Mani with his own message for the world.

Mani was born in Babylonia in the early third century AD as Persia's Sassanid dynasty (which would rule from 224 to 651) was on the make. Mani claimed to have received a vision from God and to be an apostle of Jesus: he had received this spiritual enlightenment in a cave from his own spiritual counterpart (who he called his 'divine twin'). Mani taught that Jesus was the prophet of the Jews (but had no physical body when he was on earth), that Zoroaster was the prophet of Iran and that Buddha was the prophet of India. He claimed that his church was superior to the Christian, Buddhist and Zoroastrian churches since it could transcend cultures. 'My gospel shall touch every country,' he taught; indeed, Manichaeism has sometimes been dubbed the first world religion due to the speed and distance of its spread.

Although Mani took stories and theology from Jewish Apocrypha and the Gnostic gospels, his core belief was that the universe was locked in a struggle of two equal powers and that mankind was the battleground, a dualism probably plucked from Persian Zoroastrianism. Mani believed that the good, spiritual particles of the universe had been entrapped by the evil, material particles, and that the good particles therefore needed release. The divine spirit (or *nous*) educated mankind and offered redemption, and all the great teachers, from Adam to Abraham to the twelve apostles to the three great founders of religions – Buddha, Zoroaster and Jesus – were all manifestations of this *nous*. But Mani himself was the climax of this chain of revelation, since he had received his envisioned teaching directly from Jesus the Light.

Beneath Mani, his followers were divided into the core Elect, who lived in extreme asceticism, and the Hearers, who were allowed to live more normal, materially fulfilling lives. The Elect would ascend directly to heaven at death, the Hearers would roll through a cycle of rebirth (in the form of fruit, which was considered especially holy) and the rest, those whose souls had not been awakened by the divine *nous*, would reincarnate in the souls of beasts before facing eternal damnation. The Elect had three 'seals': on the mouth (they could not lie or eat meat), on the hands (they could not kill humans or animals) and on the breast (they could not do 'works of the flesh', which included not only sexual activity but picking fruit or harvesting plants). If they kept these 'seals' intact, the Elect could release captive light through their digestive systems as they refined the particles in their food. These particles were released as they said hymns and belched at the end of meals; digestion was akin to the Eucharist.

The restriction on physical labour or earning a living gave the Elect time that they would never fill with eating, praying and sleeping. There was one obviously useful practice which remained open to such men and that was writing, the most holy pursuit of Manichaeism. A scribe might easily have spent a whole day writing just one page.

Mani saw papyrus and alphabets as the way to reach the world. He argued that his religion was superior to the other three great religions because, unlike the teachings of Zoroaster, Buddha or Jesus, the founder himself had written its scriptures down, taking their content from his own divine twin. A Middle Persian Manichaean text found at Turfan in north-west China – its content attributed to Mani – makes this clear.

> The revelation of the two teachings and my living scriptures, my wisdom and knowledge, are broader and better than in previous religions.

But he chose imagery and stories over reasoning as his religion's engine of growth, and he used the page as its porter.

Mani wrote his scriptures down to ensure they could not be distorted. His works had seven books: a gospel; a *Treasure of Life*; a summary of Manichaeism written for the king of Persia; a *Book of Mysteries*; the *Book of Giants* (based on a Jewish fable set before the Great Flood); a book of psalms; and a book of prayers.[50] He borrowed from Christian ideas, such as the apostle Paul's man of flesh and man of the spirit, and from Christian gospel harmonies, but turned them to his own Gnostic theologies. Mani also borrowed from the Tanakh (Hebrew scriptures), even as he argued against them.

Scribes were as much the propagators of Mani's religion as preachers. Mani himself did not coin their script. That had been thought to be a variation on the east Mediterranean Syriac script used in late antique Christian texts, but the dating does not tally. One possibility is that the Manichaean script came from the script of the ancient city of Palmyra in Syria.[51] Since Palmyra was destroyed in the mid-270s, that would place the emergence of Mani's script no later than the beginning of that decade. This phonetic 'Manichaean' script was used to write out his scriptures in the languages of Iran and Central Asia. The divine word thus had its own 'Manichaean' shapes on the page, whatever the tongue, delivering to the religion both a distinct written identity and an ability to cross language barriers on the page.

Love for the word propelled the scribes of Manichaeism to the top of the whole religion. If a Manichaean's calligraphy was beautiful, it was believed it was because his soul was orderly and, ultimately, divine. Manichaeism's scribes were God's own preachers. The pages they wrote brought God to the reader and thus needed to be beautifully written and decorated, as well ordered and bright as goodness itself, the mirror of man's inner spark which was, Mani taught, divine too. A well-penned page could reveal that spark and turn a reader towards God.

A Manichaean scribe had to beg forgiveness for neglecting his calligraphy, for using a damaged stylus, brush, writing board, or sheet of papyrus, silk or (in time) paper. Even the weapons of the scribe were sacred. One Sogdian Manichaean scribe recorded the words:

> If I was lazy in my writing, disdaining it or neglecting it . . . forgive me all of this.[52]

Although the focus was copying, it was not simply to accrue merit through disciplined replication, as in Buddhism, but to capture divine beauty and light in paint and ink. One Manichaean painting that survives shows rows of the Elect sitting under a tree, pens in hand. Painted around the ninth century and discovered in Gaochang (now a sand-coloured ghost town in the Xinjiang desert), the fragment contains bold, beautiful blocks of colour, and features the bowed heads of several seated scribes, some of them with pens in both hands. Flanking them emblematically are large bunches of grapes, an especially sacred fruit for Manichaeans, and the overall impression is of high holiness in paint.

As they grew increasingly expert, Manichaean scribes shrank their fonts to dwarf dimensions. In one prayer book that measured 2½ inches by 1 inch, a scribe managed to fit eighteen lines of clear letters on each page. A Manichaean book from north-west China squeezed nineteen lines into a height of 2 inches. The Cologne Mani Codex from fifth-century Egypt (written in Greek) is one of the smallest books from antiquity. It measures just 1¾ inches by 1¼ inches yet most pages carry and incredible twenty-three lines.

Often sections of a text were alternately written in red and black ink, or red was used to introduce a passage and to close it. The German archaeologist Albert von le Coq, who (having raced Aurel Stein to Dunhuang) took the best surviving Manichaean documents from north-west China back to Berlin in the early twentieth century, wrote in his diaries that the Manichaeans could not be satisfied with a text simply written in black ink. Instead they would fill their titles with colour, spread them across several pages, embellish them with flowers and other decorations, use colour contrasts and, occasionally, alternate the black lines of text with lines of text written in different colours.

Yet the beauty of Manichaean texts was not simply a matter of calligraphy. Indeed, even calligraphy itself might sometimes spill naturally into motifs, decorations and paintings, or else flowers and patterns might frame decorative titles like arabesques. Philip Larkin wrote that if he were to found a religion he would make use of water, but Mani used colour, turning book manufacturing (from papermaking through to lettering and final illustrations) into an art form that could hook the eye in at first sight. No founder of a multicultural religion had ever harnessed aesthetics to his message in this way, and scholars both Christian and Islamic complained about the stunning magnetism that the books of Manichaeism enjoyed as objects of beauty and as supposed symptoms of Mani's contact with the divine inner spark.

Mani created a book to illustrate the words of his *Living Gospel*. In Persia, he was known as 'Mani the Painter' and the Persian records of his life report that he sent a book-illustrator to accompany his missionaries on their journeys through the Middle East, Persia and Central Asia. Mani himself was probably Parthian by culture, named after the empire based in north-eastern Iran, and painting was used a little in Parthian writings, just as it was in Jewish scriptures. He surely would have seen these, as well as other decorated scrolls, in his Mesopotamian homeland.

But finding the roots of Mani's emphasis on scriptures and, more particularly, his focus on beautifully presented, written, bound and ornamented scriptures, is no simple task. So few Manichaean writings have survived that decay and destruction have made an orphan of the religion. The only extant early Manichaean documents are from Egypt and from Turfan. Manichaean groups were scattered all the way across Central Asia, Persia and the Middle East but those lands have thrown up no records. Moreover, all the Manichaean writings that survive are religious in content. We do not possess descriptions of their way of life.

The problem is made worse by the fact that the ink generally used in Manichaean writings lacks metal, which usually leaves a trace long after the visible ink fades or is washed or wiped away.

In short, it is hard to unearth the bookish roots of Manichaeism and to get an accurate picture of the religious and manuscript culture of the Asia in which it emerged. While this is not a unique problem for religions of the age, it is especially pronounced for Manichaeism due to its limited written legacy. Late antiquity was a time of extraordinary religious variety across Western and Central Asia, as Buddhism, Judaism, Mithraism, Manichaeism, Christianity, Gnosticism, Zoroastrianism and, soon afterwards, Islam were all widely preached and practised, as well as several less religious philosophies. Yet the relationships between these religions are often difficult to delineate; there is evidence of interaction but not enough to prove direct and tangible cause and effect. If each religion were a cog attached to the others, then Manichaeism was probably interlinked with them all. Yet each cog is missing some teeth too – how the religions influenced one another remains unclear.

The books themselves provide some clues about the religion's greatest influences. Manichaean texts from Egypt were written on papyrus until the fifth century, whereas the Middle East in general had made the shift from papyrus to parchment in the fourth. This anomaly leaves us without clear answers, but Mani always aimed at elites and he may have viewed papyrus as more closely associated with ancient knowledge and with reverence. A more practical reason could have been that Manichaeans were not supposed to kill animals, and parchment is made from animal skins. More prosaically, Egypt, of course, was the home of papyrus.

The Manichaean documents found in Turfan, on the other hand, are Chinese papers, and they tend to get thicker and coarser the later they were made, which is true more generally of the caches of documents found in Central Asia. This suggests that eastern Manichaeism was part of a greater manuscript culture whose hub was China. The Persians later referred to Mani as 'the painter from China', as if they assumed that this bookish culture must come from China. Less likely is that Manichaeans used brushes like the Chinese. While some of the manuscripts might have been brushed, it is hard to prove and there is not generally the variance in ink intensity usually found in longer brush strokes. Certainly, 'China grass' (ramie nettle) paper and, less often, Chinese hemp paper, were used for Manichaean books. These Chinese roots, if that is what they are, should come as no great surprise – they would simply reflect the westward progress of paper culture, before it had stepped fully out of China's geopolitical reach.

Yet whereas Buddhism in Central Asia and China often used not only the back of sheets of paper but sometimes reused sides already written on by scratching, scraping, washing or rubbing off the old text as effectively as possible (sheets now known as palimpsests), Manichaean scribes were apparently far less willing to use the same paper twice. Such reluctance is surely a reflection of the Manichaeans' exceptional reverence for written scriptures.

In the Manichaeism of Egypt and Mesopotamia, local influence is more apparent. Thus Syriac Christian hymns, or at least Syriac 'heresy', enjoyed considerable influence on the content of Manichaean writings, from the earliest heretical hymns of the second century AD – the work of a Gnostic poet-philosopher called Bardaisan – through to the use of hymns by the Church Fathers, notably Ephrem, who in the fourth century felt obliged to adopt the very melodies hitherto forbidden by the official church, so as to win people back.

Yet if western Manichaeism borrowed content from some of the same sources as Syriac Christianity was using (sources such as Bardaisan), it borrowed its book formats much more directly. The oldest Manichaean books are Greek and Coptic texts from the fifth century and they constitute some of the largest books to have survived from antiquity – while papyrus had size limitations, parchment did not. (These examples are close to A4 in size.) Like Syriac documents, these western Manichaean texts tended to use two or three columns. Over time and further to the east, fewer columns became more popular, but the initial influence was clear. More columns meant more empty space and therefore a more elegant page. However, such experiments with layout were better suited to paper than to parchment, since it could be cut to be longer, wider and thinner than parchment, as well as being far cheaper. Thus western book designs in Manichaeism could potentially be best realized on the surface of eastern Manichaeism – namely, paper.

On one torn Manichaean page from eleventh-century, north-west China, stored in the vaults of the Museum for Asian Art in Berlin, the picture sits at right angles to the text it flanks, a format also common in Syriac Christian scriptures. Thus, as Manichaeism explored the possibilities for artistry and decoration that parchment and then paper afforded, it looked to Syriac Christianity for many of its ideas. More specifically, the image did not need to relate to the text of the page it appeared on (which is typical of other surviving Manichaean books as well) because pictures were valued as media of communication in their own right. Images were crucial for Mani and it seems possible he drew his enthusiasm for them from his encounters with Syriac Christian scriptures and devotionals. When his *Living Gospel* accompaniment spread to China, the Chinese called it *The Great Drawing*, conceivably because it carried no text at all but certainly because its chief medium was images, not words. Mani had even recorded some of his cave revelations in pictures. He was deploying paper to work in new ways and with fresh power.

If some of its bookishness was borrowed, Manichaeism's evangelistic success came in its rapid geographical spread, and here it was Mani's own cultural ambitions that were crucial. Pictures are the one of the page's best expression of multilingual intentions but Mani was also determined to spread his scriptures in as many written tongues as he could. Mani's religion remained stateless, which suited both his continental ambitions and his ascetic, other-worldly priorities. And yet he also needed the patronage of kings and princes, and so it was in Central Asia that Manichaeism began to settle, to ally with local kings and to make a home. The appeal of Manichaean books had grown as they travelled east from Egypt, acquiring Hindu gods, Buddhist icons and, perhaps from Persia, floral motifs and paintings of flowers and gardens. Central Asia's Sogdian merchants certainly took to this multicultural message with enthusiasm and Manichaeism won local converts in Central Asia within Mani's own lifetime. Samarkand, where the pulse of the region's trade beat strongest, became a particular stronghold for the arriving faith.

Manichaeism also began to forge followings in the oasis towns of China's far north-west Tarim Basin around the fifth century, tapping into existing Buddhist communities for converts. More Manichaeans arrived in the seventh century too, blown in by the tailwind of the Arab Muslim military campaigns in Central Asia. Under the Uighurs, who had fled from their old northern capital of Karabalgasun to the Tarim Basin in the 840s, Buddhists outnumbered Manichaeans but the new government at Qocho adopted Manichaeism as its official religion. Manichaean priests would meet, often in their hundreds, at the house of one of their number at dusk to recite their scriptures together and to pronounce blessings on the ruler. At the Uighur court, the king set up centres for writing books and painting miniatures, while private patrons gave donations for illuminated books to be made. Having produced its own scriptures, translated them into foreign tongues and won converts across Asia, Manichaeism now had its own settlements and could begin to focus more attention and resources on its books. And it was in Uighur Central Asia that Manichaeism explored paper's potential for beautification in earnest.

Producing books in the region already fell into four stages. It began with the bookmakers, who measured, cut and attached the sheets of silk, parchment or paper, which was usually made from the ramie nettle or sometimes from hemp. Buddhist expansion had spiked demand for paper and the merchant Sogdians improved their papermaking as a result, aiding the Manichaeans in their endless quest for aesthetic satisfaction, while the Manichaean refusal to reuse paper increased demand still further. Large sums were spent on importing papers and inks from China.

Manichaeism's own multiculturalism was replicated not only in the scripts it used in Central Asia (Estrangela, Sogdian, Turkish runes and Chinese) but in its book formats too: some were scrolls in the traditional Chinese format, and a few were bound with a string passed through the middle of each sheet in the Indian manner. Closer to the Mediterranean, most were in the Western format, with the spine on the left: the *codex*. There are no printed Manichaean books and even surviving paintings show scriptures not simply as functional items (being preached from) but also as objects of beauty (the gilded images or scallops on their covers captured even in these tiny representations).

The Elect also carried out the second stage of the Manichaean bookmaking process: calligraphy. As the Turfan finds show, scribes might write a calligraphic header across the two upper margins of facing pages by stretching certain letters or by writing the entire header in multicoloured outline. Headers were sometimes painted in green or vermilion or blue. Scribes could use inks in six colours, use monochrome and polychrome outline-writing, include occasional calligraphic letters and draw simple motifs to set punctuation or other marks in relief. They learnt to draw poised floral motifs and they often left gaps, perhaps for commentary or, if a text was to be sung, they might split the words up into their constituent sounds or syllables.

The third phase of bookmaking was illumination and here the budget was crucial, as it was often Chinese who were employed as painters, and since both gold and lapis lazuli from Central Asia were used. Illuminators also filled out calligraphic headers to include people, animals and plants, choosing from four drawing styles: West Asian outline and painted, and Chinese outline and painted. For the standalone images, the best paper was used and a priming colour was typically applied first, contours drawn, painted over and then filled in with the relevant colours: several shades of blue, red, yellow and green. Gold leaf was often added at the end to make the page glisten.

Finally, the bookmakers bound the manuscripts. Manichaeans also used scrolls that were usually some 4 to 10 inches wide and, unrolled, stretched from between 8 and 13 feet. Their illuminated codices came in a range of sizes, from 6 inches by 3 inches to 20 inches by 12. Images that survive suggest many of these texts were bound in leather, but none of the bindings survive. Perhaps their quality and beauty made them irresistible to reuse for other books or for different purposes – or simply to sell off.

These remarkable products were meant to communicate to the reader through both their written and material contents. Thus the shape of the words, the order of the pages and the delicacy of the images and motifs were as important as the meanings of the words themselves. This, of course, was a function of its medium too. Paint could be readily applied to other common and portable writing surfaces but never with the same detail and precision as paper allowed for.

Manichaeism spread as far west as Algeria and as far east as the south coast of China. It was carried on paper, papyrus, silk and parchment,

in many languages but (in the east) often in the one script Manichaeans attributed to Mani himself, sometimes in codex form, sometimes
in an Indian leaf-book, sometimes on a traditional scroll. It acquired
stories and beliefs from different cultures, adapted them and then subjugated them to Mani's personal revelation from God. It was a religion
of artists and calligraphers, the first religion to win people not only on
the page and by the writings on the page but by the beauty of what it
created on that page too. It survived in China in secret societies in the
south-east until the sixteenth century.

Manichaeism's multiculturalism was a threat to both Christianity
and Islam in the early days of their spread and it incited anger, jealousy or criticism from many. Its most famous critic was the African
scholar St Augustine, who was a Manichaean himself for nine years
before he converted to Christianity.[53] Augustine often wrote and
debated against Manichaeism. He knew the power that the beauty of
its books exerted over many who came into contact with them. Their
bindings were, he said, expertly decorated, but he derided them as
trappings. In a letter to his friend Faustus in 400, Augustine punned
on Manichaeism's own theology of the evil body ensnaring the virtu-
ous soul in order to make his point.

> Burn all your parchments and their finely decorated bindings! Thus
> you will jettison a useless load, and your God who is confined in the
> tome will be released.[54]

For Islamic scholars, Manichaeism was a rival for the souls of peoples throughout Central Asia and Persia, a competing force which
could win converts by the beauty of its books as well as by its doctrines and ethics. Ibn Muhammad, an eleventh-century Persian
historian, wrote that the followers of al-Hallaj, a great Persian Sufi
and mystic, copied the book style of the Manichaeans by writing in
gold on Chinese paper, embellishing their books with silk and brocade
and using expensive leather for binding. For if it was the beauty of
Manichaeism's books that made the religion so attractive, then perhaps that beauty could be imitated.

Manichaeism's direct influence on Islam is unclear. If early Islamic
book culture was generally more dependent on Syriac and Hellenistic
ancestors, there were nevertheless Muslim historians and commentators

who remarked on Manichaean books, such as the tenth-century Bagh-dad scholar al-Nadim. But Mani had claimed his knowledge was exclusive and this claim tended to be less than popular among other religious groups. If Manichaeism was influential, it was surely in its material culture and forms rather than in its ideas, of which there are few living vestiges today.

Although one Tang Chinese emperor had a Manichaean church built in the Chinese capital, the religion was eventually hounded out of the mainstream in China by a government fired by anger at Uighur misbehaviour – the Chinese gradually came to view the Uighurs as rowdy, drunken and unrefined. In Europe, Manichaeism became a synonym for heresy. In Persia and Central Asia it survived longer, but the Arab conquests, the attacks of Genghis Khan and the gradual Islamic conversion of the region sounded its death knell.

Mani wrote down his teachings from the beginning so that no doc-trinal impurity could touch them and to enable them to conquer the world. But few books remain and Manichaeism now exists only in history. Nevertheless, it created out of the page a world to fill and decorate with text, image and motif, and its peculiar skills in produc-ing the finest books on earth did not pass unnoticed by either the Persians or by Muslims in the Middle East.

Religion in late antiquity followed a jigsaw geography across West-ern and Central Asia and Manichaeism was simply one piece among many. But its use of the book as an object of high art was both pio-neering and popular, as the anger and jealousy of Christian and Muslim writers attests. While its message was forgotten, the power Manichaeism discovered in beautifying its pages of scripture would be harnessed again, this time by a religion with a far larger destiny. Manichaeism had simply shown the way.

10

Building Books

. . . Death has no repose
Warmer and deeper than the Orient sand
Which hides the beauty and bright faith of those
Who make the Golden Journey to Samarkand.
From 'The Golden Journey to Samarkand', James Elroy
Flecker, 1913

The Ili Valley begins in China's far north-west corner and snakes into Kazakhstan. To go north of the valley, you climb past vineyards and cotton fields, pyramidal mounds with black tombstones slotted into their tops like slices of bread in a toaster, copper hills, haystacks, apple groves, lines of silver beech and occasional roadside fruit-sellers sitting at tables a few hundred feet from their farmhouses. Coal trucks climb the hill past shepherds guiding their flocks. As you leave the valley, the rim of clouds to the north gives way to blue-black mountains crested with snow. The fields below them are dirty green and dim yellow. Near the town of Huocheng you come to a walled garden. You walk along an apple orchard, down a path that passes a handful of sheep and a pair of cows. At the end you find a double door flanked by two taller trees.

The mausoleum stands twenty feet further on. It is almost a cube and its tall, pointed arch draws the eye. Tracing the arch's edge is a band of white writing on a turquoise background. Either side of the arch, lines of relief carving meet a bar above it to form a rectangle. In the panels above the arch's haunches, white medallions and midnight blue crosses form studded diagonal lines. Further out from the arch, to right and left, flower motifs forge vertical columns either side of a broader teal-green band with white writing set like musical notes along it.

Inside the arch, in the tympanum, a maze of lines and squares surrounds a small, oblong metal grill like a crossword puzzle. The whole face is crowned by the hump of a plain white dome that juts perhaps six feet above the façade. The first ten feet of wall above the ground have lost their colour so that only their stone form remains. Behind the building and to the right, the Heavenly Mountains are blue-black in the mid-distance, but turn to white as you follow them towards their peaks.

The mausoleum is the first sign, heading west towards China's border, of a foreign vernacular belonging to a foreign faith – the herald of a civilization that once dominated Central Asia and that focused much of its skill and attention on the paper page. The identity of the man for whom it was built – Tughlug, a fourteenth-century convert to Islam who ruled 'Moghulistan' in north-west China – is far less important than what the building represents. For the clues to the culture he had adopted lay written out on his mausoleum in Naksh, a tenth-century Arabic script. And the writing in the columns either side are in Thuluth, a cursive, eleventh-century script with curved, oblique lines in which a third of each letter slopes away as if a landslide has just begun beneath it.

This epitaph in stone is one of hundreds flecked across Muslim Central Asia. It was built in 1363 and is among the oldest Islamic mausoleums in the region. It stands as one of the eastern gateways into a new culture but also as one of the earliest signs of that new culture emerging. Seven years after its construction, a ruler called Timur would found an empire across Central Asia. In the West, he has been immortalized as Tamburlaine in Christopher Marlowe's play, and as Tamurlane in Edgar Allan Poe's epic poem.

Timur conquered and subdued Central Asia in the fourteenth century, from north-west China to the Caucasus, eastern Turkey, Mesopotamia, and as far south-east as the Indus River. He claimed to be a descendant of Genghis Khan and he wanted to rebuild the Mongol empire. But Timur's heritage was Persian, and his gift was to fuse his Mongol ambitions with Persian refinement. Across Central Asia, he erected mosques, madrasas and mausoleums. They were the expression, in brick and tile and stone, of his vision for a united empire. He often profited by fostering terror, once piling 90,000 severed heads into a pyramid outside Baghdad. But he was just as adept at turning art to his imperial ambitions.

Timurid buildings are Islamic but they make their own statement. Dotted across the sands, grasslands and baked earth of Central Asia, they form a bridge, a theological statement in stone. Their pale yellow bases fuse with the ground below, while their bulbous azure domes are welded to the sky above them. The sky was the muse of the Turks and 'turquoise' was their colour. These buildings echo their earlier belief in a sky-god, or conceivably point to the Persian god Ahura Mazda, identified with the sky itself. For Timur, the buildings were multicultural monuments. But they were more than that too; not only did they unite Turkish and Persian, they linked the new Timurid polity to the heavens themselves.

The largest Timurid dome in Central Asia forms a hood over the shrine of Ahmad Yasavi in south-east Kazakhstan. It spans 59 feet and measures some 90 feet from base to crest. A vast *iwan*, the three-walled space so common in monumental Islamic architecture, faces east from the shrine and either side of it, towers twist to form bevelled edges as they rise from an octagonal base. Two smaller *iwans* sit in the recess of the larger one and Kufic script is visible through the windows of the smallest. The script is white, with a blue outline, and it is written onto a sandy background with malachite decoration. Birds fly from brick to brick, making a din in the eerie forecourt.

Inside is Timur's 2-ton bronze cauldron, big enough for several people to sit in. It bears a script interwoven with decorative leaves. A very different script, white and rigid and tall, fills the band of sandy tiles below it. Set above the bowl is a vast stalactite (or *muqarnas*) corbel vault, with smaller vaults flanking it in two *iwans* off to the side. In the rear *iwan*, where aquamarine tiling reaches five feet above the floor flanked by a floral border and corner pillars in mosaic faience,[55] the doors still stand. Above, the weave of floral patterns finally gives way to calligraphy.

The dome and main *iwan* hold the skyline at the old city of Turkistan, sentinels of Timur's new world order. Across the region,

Timur's grand *iwans* tower like spiritual portals and their obelisks and columns shoot the sky. Few would leave such monumental legacies in Persia and none would do so in Khurasan or Transoxiana. Timur left them across all three.

The edifices are decorated throughout with verses from the Koran. Bands of writing dance like arabesques across the tiled faces of Timur's great monuments. On the Green Mosque at Balkh in northern Afghanistan, corkscrew pillars and a ribbed dome frame an inner *pishtaq* (the gateway to the *iwan*) marked with a broad band of Kufic script that forms the outline of a square. Lines flow outwards from the band to delineate geometrical shapes, as if the art has sprouted from the writing itself. On the outside of the drum, the supportive ring of stone that carries the dome above it, two layers of white script stand against an aquamarine background. One is curved and soft, like playful scribbling; the other is brittle as bones. In the Samarkand madrasa of the fifteenth-century Timurid ruler Ulugh Beg, Kufic texts in gold or blue form hoods over arches or bands across *pishtaqs*, inside recesses and along the columns that flank doorways. Some 130 miles west, wide bands of writing fill the whole width of the *iwan*'s three inner walls at Bukhara's Kalar mosque. Still more writing dominates the rear *iwan* too.

Timur's own tomb, the Gur-Emir at Samarkand, seems baroque in its lattice of gold, silver and blue, with complex *muqarnas* vaulting. On the outside of the dome's supporting drum, Kufic calligraphy in tiles wraps itself all the way round, almost overlapping with the arabesques and geometrical shapes that spread like wallpaper over the two pillars beside it. It is an appropriate resting place for a man who used writing as the vernacular of the architecture he sponsored.

But it is Samarkand's Bibi Khanum mosque, built for 10,000 worshippers, which spells out the link from these buildings to the paper page most clearly. For, while all these monuments celebrated Koranic writing, taking the words of the page and integrating them into their architectural forms, it was at Bibi Khanum that the page itself was used on a monumental scale. Timur commissioned a Koran for the building that, when open, measured four square yards and sat on an enormous stone lectern. Each page was just over seven feet by five feet. The whole Koran needed 1,600 sides or 29,000 square feet of paper, equivalent to two-thirds of an acre. Here was paper as a symbol of power, a physical text to act as the centre point of Timur's largest mosque.[56] It was not the only time he used writing on paper in this way – on one occasion Timur had a sheet of paper fifty feet long produced as a sign of his authority. Like his library, it no longer exists. His buildings, on the other hand, tell us all we need to know.

The calligraphy that criss-crosses the monuments of Central Asia is the clue to the next step in paper's journey. Today it is the buildings that draw us back through the centuries to the flowering of a region already in thrall to the written word. Their celebration of the Koran took many forms: interwoven with floral motifs, boxed into strict geometrical patterns or standing alone as a building's guardian verse over the main entrance. Yet this delight in the written word had begun on paper, not tile – on the pages of books rather than the walls of mosques and madrasas. Indeed, under Timur these written sentences spilled from the paper page onto the monuments of his empire, peppering the physical landscape with the very words of the Koran. It was a move calculated to attach religious legitimacy to the political power of the day, but it also points to just how much Islam had become a religion of the written word, an identity largely delivered to it on the paper page.

An Arabic saying has it that culture comes from the Persians – and it would spread across the Caliphate on paper. As the armies of Islam crossed the Levant and the Fertile Crescent from the Arabian Peninsula, their interest in the arts began to grow. They encountered cultures soaked in artistry and craftsmanship and began to learn from them. The Persians played a leading role.

Ibn Khaldun, the great fourteenth-century philosopher and historian from Tunis in North Africa, wrote that the founders of Arabic grammar were Persians. So too, he wrote, were the scholars of the *hadith*, or sayings of the Prophet. The great jurists, theologians and most of the Koranic commentators were Persians too. The intellectual sciences were owned by the Persians. Among Muslims in the earliest centuries of Islam, it was the Persians who preserved and ordered their knowledge most carefully. (Today's Iran is still quite rare among Islamic countries for celebrating its ancient, pre-Islamic history.) Ibn Khaldun even quoted the Prophet Muhammad as saying that, if learning were fixed in the highest heavens, the Persians would rise to it.

Persian peoples have inhabited Iran, Afghanistan and other parts of Central Asia for more than two millennia and the 'first' Persian empire, the Achaemenid empire founded in the sixth century BC, was the most powerful and magnificent kingdom of its day, spanning a region that included parts of Asia, Africa and Europe. Thus, when the Arab armies conquered Persia, they simply put on its robes of state. Iranians were spread across Asia, from Mesopotamia to Transoxiana, and they peopled the new Islamic bureaucracy, filling all corners of government. As the bureaucracy's clout subsided, imams and authors came to the fore. The offices of state were Arab but the theories that underpinned them – theories of governance and of administration – were thoroughbred Persian. In great part these theories emphasized the role of the literary and religious classes in administrative life, and the work of these classes was done on the page.

Yet Islam did not simply adopt a Persian identity. Instead, Arabs and Persians cross-fertilized ideas to create a more universal religion. Arabic remained the language of Islam but it was Persians who became its scholars. The Sassanid Persian empire ended with the Muslim conquests but its royal art was funnelled into an Islamic mould. Royal images, motifs and symbols were far enough removed from religion (and so far enough removed from heresy) to become a bank of icons for Islam too.

The early caliphs were not interested in foreign conversions. In the sixth century Caliph Umar tried to limit conversions to the Arabs. Converts meant more soldiers to divide booty between, more pensions to pay and less poll tax revenue. In Central Asia, it was especially difficult for locals to convert. On one occasion, the promise of a tax break was repealed after too many locals converted. A rebellion ensued. The first Islamic Caliphate, which ended in 750, was undone by its failure to translate Islam out of Arab culture. Its successors, on the other hand, sought converts as much as wealth; as they planted nomadic, Arab Islam in the rich soil of Iranian cultures, a new form of Islam began to take root and grow.

Iranians took Bedouin poetry from the Arabs and refashioned it to suit their own tastes, spawning new forms. The new poetry was in Arabic but its appeal was far broader. Meanwhile, many Iranians devoted themselves to studying Islam and Arabic. One mid-ninth-century Iranian religious scholar would read from the Koran, and then explain it in Arabic to the Arabs on his right and in Persian to the Persians on his left.

Iranians, already fond of learning and writing, studied in madrasas and soon became the Caliphate's best scribes. It was two Iranians, Ibn al-Muqaffa and Abd al-Hamid, who became the pioneers of Arabic literary prose. Iranians lived in Arab communities just as Arabs lived in communities of Iranians. The empire was turning bilingual but its focus was the one script, Arabic, and its glue was the one religion. In order for such a bookish religion to reach its potential, paper was essential and, by the fifteenth century, Iranian papermakers could make papers of virtually any useful size, strength or texture.

The Arabs had spread Islam under the banner of holy war and conquest, but Persians ensured it was able to acclimatize abroad. Yet a closer look at the great Iranian scholars, commentators, scientists, inventors and calligraphers of the early Muslim empire throws up a surprise. These men did not come from western Persia or Mesopotamia, where Islam first encountered the fading Persian empire, but from further afield – from eastern Persia, Khurasan, and the lands south of the Oxus River, as far south as northern Afghanistan. (Khurasan covered parts of today's Afghanistan, Iran, Turkmenistan, Uzbekistan and Tajikistan.) To the west of Persia, including Mesopotamia, most Muslims were Shi'ite whereas further east Sunni Islam, the predominant form, held sway.

The cities of Bukhara and Nishapur produced the two greatest collectors of the *hadith*, al-Bukhari and al-Hajjaj. Across Khurasan, scholars studied apace to learn pure Arabic and soon became famous for the *Koranic* style of spoken Arabic. Indeed, the ninth-century geographer al-Muqadassi said it was the Khurasanis who spoke the purest Arabic because they studied it so assiduously. The first Arabic dictionary was recorded as being written in Khurasan in the eighth century. Khurasan and Transoxiana also supplied many of the silks, furs, textiles and silver objects sold in the markets of Baghdad. Zoroastrians fled to its cities from central Persia, while its growing wealth and sophistication attracted Arab settlers too, and supported a new age of scholarship and invention.

New Persian was first saddled for use in these lands too. This fresh language adopted many Arabic words, a growth supported by the writings of Arab and Persian poets. But it crystallized for the first time in the east, where the population was more mixed. Khurasan became the factory of Islamic civilization as nowhere else could, taking on Arabic as the script of science and learning with ease, but turning Persian into the spoken language of the Caliphate's east. Iranians had not been so united since the fall of the Achaemenid empire in the fourth century BC.

This young Persian, Muslim culture suckled on other parent cultures too. Chinese painting and Manichaean texts aided the emergence of Persian miniatures. Painting schools sprouted in Herat, Shiraz and Tabriz. This new Islamic powerhouse helped to forge a more ecumenical religion, one that could convert as many people by persuasion as by force, threats or the offer of tax breaks. It spawned new warriors at the end of the ninth century too, warriors determined to expand the frontiers of Islam still further. Yet even among them, education had made their vision of Islam multicultural: a fusion of Arabic, Hellenistic and Iranian learning.

The madrasa, or Islamic school, was born in Khurasan and Transoxiana. The Koran, theology and jurisprudence all became elements of religious education under the Abbasid Caliphate (founded in 750 to replace the Umayyad Caliphate; the Battle of Talas was an early Abbasid victory). In Central Asia, Muslim jurists grew in stature; even today, they perform a vetting service that sits awkwardly with the more democratic institutions of Iran. The madrasa would gain increased independence from the mosque from around the tenth century. Soon, professional reciters were employed in towns throughout the region so that the Koran could be heard – as late as the nineteenth century, the Tajik city of Khojand alone had seventy reciters.

In the tenth century Muhammad ibn Ishaq wrote that Khurasan paper was made from linen. Already, it could compete with the best Chinese papers available. But paper's new patron was Islam, and the faith raised calligraphy to the region's highest art form. Persians used both parchment and papyrus, the former common for letters and bureaucracy. In Egypt, the Sassanid Persians had even written on the hides of cows, sheep and buffaloes. For royal letters, they had used scented silk or parchment, closing them with a seal and then placing them in multicoloured damask envelopes. A hoard of eighth-century letters was found near Mount Mug in Tajikistan by Soviet scholars in 1933. Most were written in Sogdian and were on wood, silk or skin. A few were on Chinese paper and parchment. Soon, however, Samarkand was profiting from its own paper industry.

Around the turn of the eleventh century the Baghdadi bibliophile Ibn al-Nadim recorded that the Chinese wrote onto a paper made from 'a kind of herbage'. He went on to say that this material provided a large income for Samarkand itself. He, too, argued that the Arab papermakers had initially acquired their craft from the Chinese captives in the city. The claims appeared in al-Nadim's great work, the *Fihrist*, a kind of compendium of all books which existed in the Arabic language. Such a work could hardly fail to mention the surface that had fuelled book culture so effectively. Islam reduced the Babel of scripts used across Central Asia to one above all, Arabic, and it focused attention onto one book too, the Koran. With that focus, bookmen emerged across the region and the ancillary skills of papermaking, bookbinding, illumination and calligraphy became as God's own arts.

Sheep graze among the walls and ditches of the old hill city of Marakanda, founded twenty-seven centuries ago and conquered by Alexander the Great. After the Muslim conquests, Arabs and Turks migrated to live there in numbers, and its wealth spawned artists. Seen from the ruins of Marakanda towards dusk, the city of Samarkand below you radiates a warm light. Grand *iwans* and minarets spike the cityscape. These are from the time of Timur, after the Mongols had destroyed the old city. Around it the countryside is rich, carpeted in the green that only centuries of irrigation can provide. Marakanda was always a city of gardens flanked by fields. But papermaking needs water too, and this supply, coupled with the prolific local flax and hemp, granted Marakanda a papermaking boom. Together with Herat, it became the production centre for converting Central Asia to Islam – on the page.

Its new product spread quickly. Persia was using Khurasan paper in the eighth century. In Central Asia, papermaking soon became the first step in a far longer process – to build Korans whose physical beauty reflected their identity as the words of God. Rules and standards for dimensions, illustrations, font sizes and styles, margin decorations, chapter headings and miniatures all developed as part of a vision of a harmonious and carefully composed page. The execution of this vision depended on an army of calligraphers, decorative artists, paper-makers, miniaturists and bookbinders. By the fifteenth century, Iranian papermakers were making dozens of different types of paper, varying in size, thickness, colour and quality. From the thirteenth to the six-teenth century, these papers were the best to be produced in Islamic lands.

Central Asia had found a force capable of unifying it. Buddhism preferred printing to displays of calligraphy and Manichaeism was in slow decline. Foisting Arabic onto the region as the only language of scripture placed a new emphasis on the bones of the letters them-selves. The very strokes of a brush or pen spoke as clearly as the sounds they represented.

Spacing, proportionality and visual balance became the goals of the calligrapher. But he could work with strokes that were thick or thin or that varied in width, with different spacings between letters and words and with a range of overlapping and intertwining formats. Some styles were sharp and cubic while others were fluid and curvilinear, a line of ballerinas caught on the page. It was not simply that the blank page was a vehicle for the written language. Instead, the script itself was like a blank page for the artist-calligrapher too, affording him huge scope for variation in form and mood. At times, beauty trumped legibility.

Naskh, a cursive script, was standardized in the tenth century and travelled east, finally arriving on the buildings Timur erected. (It is also the preferred script for printing in Arabic.) In the twelfth century, calligraphers often used two different cursive scripts, one for the text of the Koran and the other for a commentary. A volley of commentar-ies written across Persia adopted this double-writing; in Nishapur, the author al-Surabadi used it to pen fabulous stories about Adam, Noah and Solomon. Islam allowed Central Asia to adopt Arabic not only as a script but also as a potter's clay, receptive to the region's artistry.

Often books were signed twice, once by the calligrapher and then by the illuminator – such was the perceived importance of their respective roles. Illuminators embellished the first and last folios of a book. They designed the layout of the text and its borders. They penned a large medallion at the centre of the first folio as a billboard for the owner's name. And they drew a scalloped headpiece across the page-wide frontispiece of the second folio, to carry the book's title and the *bismillah*, the refrain which opens every chapter of the Koran but the ninth. Often the second folio showed a royal hunt or a maze of geometrical shapes. These frontispieces were the portals into the written word. Islam did not simply put text on paper: it constructed its books with an architect's eye.

Further into the books, a gold and blue frame might envelop the text while headpieces carried chapter titles, and margins were filled with patterns, plants, animals, birds, appliqué work or incrustation. Many of the techniques had come from China, notably dyeing, marbling, tinting and gold sprinkling. (European travellers took Persian marbling techniques back to Europe in the seventeenth century.) Often a gold spray was spattered across the margin too, or a blue, yellow, red or orange tinting. Framed by these margins, the words of the Prophet were laid out across the page, actors delivering their lines on a painted stage.

But there was no sense of perspective on the medieval Muslim page and no chiaroscuro either. Thus, when landscape paintings began to appear on Central Asia's pages, they pillaged Chinese styles. The Mongols arrived in the 1220s under Genghis Khan and Chinese brushwork crossed Asia as a result of their westward progress. In Persia, the cities of Shiraz and Tabriz became centres for miniature painting; further west, so too did Baghdad. By the 1420s there were forty master calligraphers in Herat alone and many more manuscript and miniature masters too.

The Mongols would split their empire four ways. Central Asia was handed to Chagatai, Genghis Khan's second son, and in the early fourteenth century the calligraphers, illuminators and bookbinders of the lands he ruled began to create Korans of a new order. One Koran given to the Sultan measured 28 by 20 inches, yet each page carried just five lines, in alternating black and gold.

In what is now north-west Afghanistan, the Herat school of miniatures used soft, light tones as background for the bold, glittering
clothes of characters in their pictures. Over time, pictures grew more
detailed, a warren of exactness without focus across each tableau.
Moreover, large sums were spent on books produced in Herat, and
even on the paper. One elaborate fifteenth-century copy of the
Shahnama, an epic eleventh-century poem on the history of the Persian kings, cost 42,450 dinars, an enormous sum, and was worked on
by calligraphers, illuminators, margin-makers and painters; 12,000
dinars of the total were spent on the Chinese paper.

Then, in the late fourteenth century, Timur began to reel scribes and
artists back to where the region's papermaking had begun, his new
capital at Samarkand. A fresh, fluid and curvilinear script called *nast-
aliq* was invented and became the vehicle for Persian poetry. In
Samarkand, paintings took on a spare format, less crowded with tan-
gential detail. Meanwhile, in Herat, calligraphers began to exaggerate
the turns and flicks of Arabic letters, thickening them or cutting their
reed pens at an oblique angle. (Fountain pens were invented in Egypt
in the tenth century but were not popular in Persia or Central Asia.)
They even used découpage, cutting letters from paper to stick on
another sheet as text.

This obsession with paper, scripts, bookmaking and book decoration helped Islam to win the region for the new faith. There were
forced conversions, tax and job incentives too, but it took years to
evangelize the countryside and to turn Islam from an Arab import
into a world religion. The scholars of eastern Persia and Khurasan,
however, embedded Islam locally, and their calligraphy, papermaking,
illumination and bookbinding were the nuts and bolts of the change,
since they beautified the books of Islam with local colour and decoration. Thanks to their work, Koranic texts and commentaries became
not only more widespread but also the focus for high culture in the
region, *objets d'art* that adorned the new faith and won admirers
through their physical beauty as well as their written content.

Yet the evidence of the Timurid love for the written word, and for
the beautified word, survives today largely on stone, not paper. In
fact, buildings *were* the books that forged an imperial landscape in
Timurid Central Asia, introducing all sorts of novelties, including the

squinch (the Persian corner hood that allowed the transfer of weight from a dome to a square base). The old city of Samarkand stands today as Timur's showpiece, an array of ribbed domes, geometrical designs, arabesques, minarets, verses and patterns written in brick and tile, *iwans*, portals, *mihrabs* (the recesses that face Mecca), stalactite vaulting, courtyards, *pishtaq* fronts, silver inlay, relief carving, lapis lazuli, gold and the letters of Islamic calligraphy, Arabic in tile and stone. Many of these features had been designed on paper first – in other cases, they were really just the overflow of decorations already in use on the page. Such influences would continue to be felt further afield, after Timur's grandson Babur founded the Mughal dynasty, its great buildings, dotted across Rajasthan, built on the traditions established in Central Asia. Yet it is the city of Herat which claimed the finest expression of the Koran in stone, brick and tile.

Herat spreads in tree-lined streets from the Friday Mosque to the fields. In the streets that flank the mosque, shops sell fabrics, gravestones, carpets, burqas, clothes and rifles. Stalls along the pavement sell shoes, shampoo and fruit smoothies. The Mausoleum of Gohar Shad, Persia's great female patron of the arts, sits in a dry and mournful garden at the edge of the city. The dome has lost its colour but still shows signs of its antique glory: double-octagon squinches carry the dome's weight to a square base. Five of the nine minarets remain, slanted and crumbling. (The other four fell as Afghans prepared to fight against the Persians and Russians in 1885, helped by a lone British general acting against orders.)

It was Gohar Shad who moved the Timurid capital from Samarkand to Herat in 1405 and gave the Persian language new sway. Today Herat remains a city caught between Persia and Central Asia; for a century and a half its largely Persian citizens have been under Afghan rule. This rich confusion is played out not only in the faces and facial hair of its residents but in the building at its centre, too, the Friday Mosque.

Its courtyard is paved with long, white flagstones. These, together with the arches, add serenity to offset the detail of the panelling on *pishtaqs* and *iwans*. Inside the courtyard, the *iwan* sits only twice as high as the arcades that flank it, three archways wide on either side.

Plain marble base panels stand five feet tall, while archway panels and pillars form a mosaic of patterns: leaves and branches, geometrical shapes and medallions, like a pointillist painting using only blues, greens and occasional yellows. A band of texts five feet tall spans each arcade, its letters running rhythmically across them like the electrified forms of a Jackson Pollock, with white writing on a background of blues. More writing fills the pillars and spans the main *iwan* and soft yellow Kufic script dances on the western *iwan* overhead. Everywhere art and calligraphy intertwine across a monument whose decorations support and embellish one another. The result is one of the most heavenly buildings on earth.

13 The Friday Mosque, Herat. It was here that Timur's monumentalism was most successfully subdued by a Persian aesthetic. The form is contained and simple, enabling the building to hover rather than soar, and for the eye to be drawn to the elegant archways and to the unending foliated patterns of the upper panels. Disciplines of the paper page were crucial for the erection and decoration of such buildings, among them geometry.

The Friday Mosque shimmers turquoise in the midday heat, graceful and intricately patterned. But the rows of calligraphy painted across its panels are a reminder that the building forms part of a broader civilization spanning Eurasia, a civilization founded on a single book. In the story of paper, these monuments act as the most arresting relic of an age it might otherwise be hard to access or visualize. The mosques and madrasas of Central Asia are not so much the next step in the paper trail, as they were largely built centuries after the initial transfer of papermaking to the region. Instead, they are

perhaps the finest flowering of paper's second great partner civilization – a museum of exhibits which point back to an age when paper
revolutionized the region. Paper provided the surface not only for the
Koranic verses that now decorate walls and arches, but also for the
drawings and mathematics which made such buildings possible, and
for the planning of those geometrical patterns and arabesques which
help to make them among the finest monuments in the world.

In Timurid Central Asia, writing, geometry and art all spilled over
from the page onto the monuments that still stand to this day. Paper,
cheaper than all predecessors and far more versatile, allowed calligraphers and artists to experiment more freely, and to focus not
merely on the message of a book but also on its form, decoration
and image. Most of these experiments are now lost to time and
decay, but the decision to transfer their results to the empire's finest
buildings means that their legacy remains strikingly visible. These
monuments are crucial evidence that the progress of paper from
western China to the Mediterranean is essentially an Islamic tale. It
is therefore not possible to trace that progress without first turning
back in time to the one book that brought it all about, to the man
who was its author and to the scholars and rulers who finalized its
contents.

It began life as an unwritten recitation performed in a proudly oral
culture. And yet the Koran gave birth to a civilization that spanned
Eurasia, one which placed books, and therefore paper, at its very
heart, propelling literacy down the social scale, advancing arts and
sciences to unparalleled levels and setting in place a vast paper bureaucracy as its tool of government. No single book had ever spurred the
advancement of a highly bookish culture on a continental scale. Yet
this had come from an author who felt uneasy about setting his teachings down in writing in the first place.

In hindsight, Islam's partnership with paper *can* appear almost
inevitable. Paper allowed the Koran to travel the world and to embed
itself in cities as distant as Cordoba in Spain and Delhi in India. Any
faith might win new lands through conquest in the short term. But to
last down the centuries, some sort of unchanging authority was
needed and the Koran provided just that. It is perhaps not hard to see

how instrumental paper has been in the success of Islam and of its book of revelation, the Koran.

Yet that is to look at it with our own, book-wearied eyes. Sixth-century Arabia, where Muhammad was born, had no great attachment to books. It was predominantly an oral culture, one which revelled in recited poetry as well as in a life of wandering. It is no great surprise that Muhammad himself – who may have been illiterate – felt uneasy about the use of writing. Had that coolness towards written texts persisted, paper's journey would have looked very different; it would certainly have taken far longer to make its mark beyond the lands of Greater China.

In China, writing had been a pillar of the civilization for several centuries before the arrival of paper. Thus, once the Chinese literati had accepted a lower-brow writing surface than the silk and bamboo they were used to, paper's status was secure. But an Arabian religion was a far less obvious partner for paper's global progress. It is easy, looking back, to miss just how unlikely it all was.

II

A New Music

*He entrusted knowledge to a piece of paper, thus squandering
it. Woe to the man who entrusts knowledge to mere paper.*
                    Ibn Abd al-Bar, tenth century

It all began, or so we are told, when Zaid ibn Thabit's master ordered
him to scratch out a few consonants onto the shoulder blade of a
camel.

Zaid's master, after all, was probably illiterate. Their home was a
merchant town in an oasis that lay on the trade route from Yemen to
Syria, Iraq and Palestine. By the time of Zaid, who lived in the seventh
century, the camel had emerged as Western Asia's most popular beast
of burden and had supplanted even the wheel. Zaid's people, the
Arabs, owned the camels and increasingly controlled the trade flows.
But his master, Muhammad Ibn Abdullah, had lost interest in pursuing trade as a career. As he observed oasis life and the rise of a new
breed of merchants, he watched them discard their old social values,
and he objected to the injustices and selfishness that ensued.

Muhammad was born around 570 and, as he grew older, he became
convinced that God would not forget the peoples of Arabia. His own
tribe, the Quraysh, was responsible for upkeep of the Ka'aba, the
cuboid granite home of the 'Black Stone' in Mecca to which devotees
made pilgrimages once a year from all around the peninsula. The
Ka'aba was meant to be where heaven and earth intersected and
where the original house of worship had been built by Abraham. The
'Black Stone', which today sits in one corner of the Ka'aba, is reckoned to survive from early antiquity. The Ka'aba had certainly existed

as a place of pilgrimage for centuries before Muhammad was born, the 'Black Stone' all that now remains of its oldest elements, and had played home to many idols. By Muhammad's time, however, it was treated as a sanctuary for the one supreme God. And yet this affiliation was not enough to satisfy Muhammad, who knew Jewish communities and probably met wandering Christian monks and preachers too. These peoples had their own prophet; so did the Persians in the person of Zoroaster. Why not the Arabs?

Like his contemporaries, Muhammad visited the Ka'aba, but he also liked to withdraw to the cave of Hira, close to Mecca, to meditate and to escape the city. One such visit was to upend his life, his tribe and, ultimately, the whole Arabian Peninsula. For in the cave Muhammad heard a voice.

'Recite!'

'What shall I recite?' he replied.

'Recite: in the name of your Lord who created, created man from a blood-clot. Recite: And your Lord is most generous, who taught man by the pen, taught man what he knew not.'

Muhammad was terrified and hurried back to his wife, Khadijah.

Khadijah had first hired Muhammad, fifteen years her junior, as an agent to travel with her trade caravans. She was so impressed by the returns he managed and by his reputation that she decided to marry him. When he returned from the cave to his wife's side, Muhammad was not sure whether the voice had truly come from God. But Khadijah consoled him, reminding him that he was a good kinsman, generous to guests, supportive of the weak and kind to the destitute. The message must be God's, she assured Muhammad. Khadijah is remembered as the first Muslim.

Thus began Muhammad's revelations, which continued throughout his life. Three years after his first cave revelation he began to preach his new message and to draw a following. According to Islamic tradition, some of his revelations would be written down quickly, on the bones of camels, sheep and asses, on potsherds and white stones and leather, on date palm leaves, parchment, papyrus and wooden boards. Islamic historians have traditionally depicted Muhammad as illiterate, adding to the miraculous nature of Koranic revelation. Muhammad certainly began to employ scribes, but the Koran mostly began its life as a growing body of sermons for his Meccan followers to live by and recite. In this era, many Arabs still distrusted writing: their poets were lovers and masters of the spoken word. Indeed, when they gathered at sites across the peninsula each year to make religious observance, poetry recitation competitions were held. Traditional sources place Muhammad at these events, but not as a participant.

Even today, most Muslims encounter the Koran first and foremost as a recitation rather than a paginated script. Life's most important moments are marked by Koranic readings and the written Koran is more a musical score to be performed than a book to be silently read in private. It is recitation that allows the voice of the Prophet to re-appear. Muhammad was slow to accept the writing down of his personal sayings, but his Koran became the one of seventh century's largest anthologies of prose-poems, and a book that drew on several sensations at once: sight, touch, speaking and hearing.

Muhammad's revelations spawned an empire of writing which stretched from Central Asia to Spain. Before the West began to print books, the Koran was exceptional in its geographical reach and the size of its audience; in modern times every home and every Muslim would own a copy. It would also launch a dozen related scholarly disciplines, such as lexicography and grammar. The Koran's destiny was the paper page and, in time, paper would become the administrative currency of the Islamic empire, an empire as large as any that had gone before it. It had come from mercantile Arabs, who raised up the Koran as the lodestone of a new civilization, forging a community around it.

The Koran's impact on the story of paper is therefore exceptional. It was born as a recitation in a largely illiterate community of urban and nomadic Arabs, who had previously preferred the oral performance of poetry to the written page. Yet it was these Arabs who took that recitation to the page, who raised up the Koran as the lodestone of a new civilization, forging a community around it and, in so doing, raised the paper book to a new status. Books were placed at the centre of Arab and Islamic culture and life. Its rulers and civil servants brought that book culture into all the lands under Islamic rule, which reached as far as the borders of China to the east and the Maghrib to the west, later spreading north into Europe. A culture of reading (and of written bureaucracy, as imperial governance took shape) was therefore introduced to peoples thousands of miles apart and, from the ninth century, it took place on paper. This movement, oceanic in its scale, began with the written Koran.

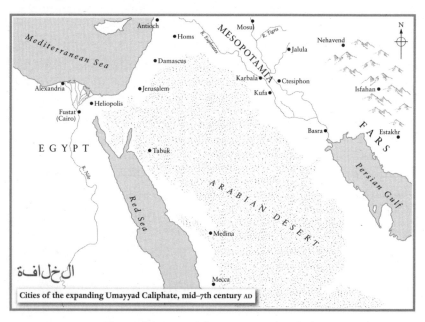

Cities of the expanding Umayyad Caliphate, mid–7th century AD

14 Cities of the expanding Umayyad Caliphate,
mid-to-late-seventh century AD.

Muhammad moved to Medina (then called Yathrib) in 622 and succeeded in uniting his immigrant fellow tribesmen with their host community to form a single community of the faithful. All the while he continued to pass on his revelations, which Koran 3:103 described as 'the rope of God'. In their view, Muhammad was no more than a mouthpiece, receiving God's words directly from the angel Gabriel.

This is a crucial point for understanding why Muhammad's new teaching became paper's great ally. In the story of paper, one of the greatest surprises is how a religion of a fractured, tribal society, many if not most of its members functionally illiterate, came to rest entirely on a single book and the study of that single book. The process that places literacy at the heart of a community's way of life for the first time is rarely simple. But it is harder still to navigate that process when literacy and illiteracy sit side by side, when religious influences are both external and internal, and when a community has its own language but is also familiar with non-local dialects.

Prior to the Koran, there were only the beginnings of significant Arab writings to draw on, but there are plenty of echoes and parallels of non-Arab cultures to be found in the Koran. Some scholars have seen strains of the Jewish Exodus story in the Islamic *hijra*, the migration of Muhammad and his followers to Medina, and Islam's declaration of faith, 'There is no God but God', resembles the Samaritans' earlier liturgical refrain 'There is no God but the one'.[57] The Koran's claim that Jesus never physically died on the cross due to a last-minute substitution echoes the teachings of Docetism, an early Christian heresy. Some of the Koran's stories also appear in the earlier Jewish Midrash and in Jewish Apocrypha. The Koran similarly appears to draw heavily on stories from Genesis, the gospels and Christian Apocrypha. Sidney H. Griffith, a leading historian of the period, has presented evidence that the Bible, originally penned in Hebrew, Aramaic and Greek, was not translated into Arabic before the rise of Islam.[58] There is a much stronger case for religious influence from Syriac and many of the Koran's verses, the more controversial Syriac philologist Christoph Luxenburg[59] had argued, closely mirror fifth- and sixth-century Syriac Christian hymns. This should come as no surprise, however: scriptures are typically part of a religious conversation across beliefs and cultures. The Koran itself even makes passing reference to other religious works, such as the Hebrew Scriptures: 'to David we gave the Psalms . . .' (4:163). The Koran assumes its audience is familiar with the Jewish and Christian scriptures. Gabriel, an angel in the Christian scriptures, is never identified as an angel in the Koran, but is clearly still understood to be one.

Thus any religious text is in some sense responsive. But concepts of the Koran being influenced by pre-Islamic human agency, even in the form of scriptures, have traditionally been unwelcome within Islamic scholarship. The idea of the Koran having any kind of decisive human or literary context is problematic to the traditional view of the Koran, which takes the book as a standalone revelation from God. This has fettered the study of the Koran's origins.

Nevertheless, in the 1930s two Western scholars, Arthur Jeffery, an Australian, and Gotthelf Bergsträsser, a German, decided to write a critical history of the Koran, both to chronicle the text's creation and to amass all references to its variant readings from across the whole of Arabic literature. They travelled to see the oldest Koranic manuscripts and returned to Germany with perhaps 15,000 photographs of early Korans and of early references to variant readings. But Bergsträsser, a professor of Semitic Studies and an eager opponent of the Nazis, died on a mountaintop in 1933 in strange circumstances. (One Egyptian academic openly blamed the Nazis for Bergsträsser's death.)

Another scholar, Otto Pretzl, continued the project with Jeffery, and much of the work was finished when he was killed in fighting outside Sebastopol in 1941. After the war, in 1946, Jeffery explained to a crowd in a Jerusalem lecture theatre that the whole photo-archive in Munich had been destroyed by bomb action and fire. He concluded that a truly critical edition was unlikely within a generation. Jeffery died in 1959. In the 1970s Dr Anton Spitaler, a professor of Semitic Philology at Munich University, confirmed that the collection had been destroyed. (Indeed, it was probably Spitaler who had informed Jeffrey of the archive's demise.)

There, it seemed, the story must end, until fresh scholars could access all those materials once again, with all the funding and political will that such access would require. Yet, after Spitaler retired in the 1980s, it emerged that the 450 rolls of film had in fact survived and that Spitaler had passed them on to a former student. It is these photos that are now under scrutiny at *Corpus Coranicum*, an eighteen-year research project at the Berlin-Brandenburg Academy of Sciences. The project is cataloguing photographs of the oldest Korans from around the world as a public research archive, but a key element in the project is focused on intertextuality: the relationship between the Koran and other texts that may have influenced it. Michael Marx, who runs the project on a day-to-day basis, views it as an attempt to begin to map the Koran in its earliest days. 'We are hoping to rein in that untamed beast, Koranic scholarship,' Marx told me. 'We want to show the textual history of the Koran, just as has been done with the Bible, with Shakespeare and with Goethe.'

There are plenty of clues that the Koran had a dynamic relationship with other faiths, cultures and texts. There were already pilgrimages to Mecca before Islam, just as there was, according to Islamic tradition, a Christian icon in the Ka'aba, long since Islam's holiest site.

The Koran accepted some elements of pre-existing faiths but self-consciously rejected others. Thus, sometimes it agrees with Jewish and Christian scriptures, but at other times it integrates elements from Genesis and the gospels not to rehash old arguments but to set itself up in opposition to them. In one story from the Infancy Gospel of Thomas, a second-century apocryphal gospel, Jesus is described as making birds from mud and giving them life, after which they fly away, a story which suggests his divinity even in childhood (a claim some devotees denied). Yet the Koran, which rejects the divinity of Jesus, chooses to use that same story, adding simply that Jesus is only able to give life to the birds because God grants him that power in that particular moment – the lesson had been deliberately reversed.

Elsewhere, the Koran takes a well-known theological formula and turns its meaning on its head, reinterpreting the Syriac Christian version. It affirms, for example, that 'God is one, eternal, has not procreated nor was he born ...' (112:1–3), but this teaching is also the basis on which it discounts the divinity of Jesus and, further, the possibility of God choosing to become a man. Moreover, in terms of its literary form, the Koran is not the heir of either the Tanakh or the gospels and letters of the Christian New Testament – it is a very different form. For all it shares with Christianity, the Koran is not a mere recalibration of Christian theology. Michael Marx believes the historical freshness of Islam can sometimes be lost in comparisons. 'Islam is not apocryphal gospel number 99,' explained Marx. 'It is something new. There is a prophet who has a different idea and, against the backdrop of church history and all these late antique cults, he starts installing a new theology.'

There was much in that backdrop to respond to – not only the thriving Syriac Christian communities of eastern Turkey, Syria, Mesopotamia and Arabia, but also the large Jewish communities in Arabia. There were pagan cults, Manichaean communities and still others.

Yemen itself was nominally Jewish, and surviving inscriptions make clear that the Yemenite rulers accepted monotheism, though they often leave out details associated with Jewish religious practice. It was in this context that Arab rulers, aware of Syriac Christianity and its many writings, began to channel their own religious identity, which lacked some of the defining characteristics of Christianity and Judaism, such as a well-worked doctrine of God or an accepted set of scriptures. In such a setting, it is perhaps easy to imagine Christians and Jews asking Muhammad's new followers why they have no scriptures of their own.

The Koran's relationships with pre-existing texts, then, are enormously important to the rise of paper, since they help to answer the question of how such an avowedly oral culture could have become one of paper's greatest allies, spurring paper's spread across Asia and North Africa into Europe. But these influences remain difficult to pin down. Indeed, the beginnings of the Koran are more disputed among historians and Koranic scholars than agreed on. It is not even clear that the Koran is rooted purely in the tradition of oral recitation, as Islamic tradition has asserted. The words 'quran' (recitation or liturgy) and 'kitab' (book) both appear regularly through the 114 suras of the Koran, suggesting some level of acceptance of both forms. Certain words point to the importance of written sources in the compilation of the Koran too. 'Furqan' is used in the Koran to mean both 'salvation' and 'commandment', giving the word two derivations: 'purqana' is Syriac/Aramaic for salvation, 'puqdana' is Syriac for commandment. Yet while these two words sound very different, they look very similar. Conflation into the single word 'furqan' therefore suggests the use of physical texts in the process of transmission.[60]

Muslim dating begins with the *hijra* of 622, when Muhammad and his followers left Mecca for Medina, and there is no Arabic term for 'pre-*hijra*', no equivalent of 'BC', and therefore no nod towards older sources. As for form, early copies of the Koran were not scrolls, like the Jewish scriptures; they were codices, the book form already used by Christians. But their script, running from right to left, distinguished them from Christian scriptures, and soon they developed their codices to be more horizontal, further marking out their uniqueness.

*

The Koran's official compilation is better documented. Islamic tradition places the death of Muhammad in 632 (although at least one source places it later). By this time, the west and south of the Arabian peninsula recognized Muhammad as God's Prophet and ruler. But the death of the Prophet brought a new challenge. He could no longer pass on or confirm the revelations he had received and there was no library of his teachings. That same year, Islamic tradition narrates that hundreds of the great Koranic reciters were killed at the Battle of Yamama, along with thousands of eastern Arabians, who had been living among the palm orchards of their desert oases.

The death of so many reciters vexed Abu Bakr, Muhammad's successor as caliph, since there was a real danger that Muhammad's revelations might actually end up forgotten or only half-remembered. So Abu Bakr ordered a gathering of the verses of the Koran, a collation of the words of God, whether they were on parchment, on bones, on date palm leaves or in the memories of the people. Although they were to be checked against oral tradition, this was nonetheless a textual project.

It was Muhammad's son-in-law Uthman, who came to power in 644, who is remembered as the father of the physical Koran. Medieval Muslim sources sometimes call Uthman the editor, sometimes the collector, sometimes simply the copyists' boss and sometimes the binder of the Koran. Uthman, was, above all, the canonizer.

It is fortunate he was a good diplomat too, for an argument soon broke out in the city of Homs in Syria between Syrian and Iraqi Muslims over their different versions of the Koran. (Different major cities in the Caliphate housed slightly different versions of the Koran within their own walls; each *mushaf*, or codex, was based on the reading of one the Prophet's own companions. Written differences between them were often only in such areas as phraseology and sentence structure.) One of Uthman's generals accused both parties of squabbling like their pagan ancestors, left for Medina and, ignoring tribal protocols, went straight to see Uthman. He told the caliph that the different versions of the Koran needed to be reduced to one or else Muslims would get caught in an argument as untidy as the one between the Jews and the Christians.

Islamic tradition recounts that Uthman appointed Zaid ibn Thabit,

that self-same scribe who had sat with Muhammad etching consonants onto a camel's shoulder blade, as editor-in-chief for his new
project, a single and standardized Koran. Zaid was helped by three
further editors and by those of the Prophet's Companions who had
survived. (All 'Companions' had a personal link to Muhammad and
were therefore believed to hold exceptional authority.) Zaid was the
best copyist and he was aided by a linguistically proficient Companion, who dictated all the texts Uthman had gathered. This was the
moment that written text truly began to compete with the oral tradition in Arab-Muslim culture, not least because the project relied on
the assumption that the page (at this stage still parchment) was the
best available guardian of accuracy.

Uthman was thorough, ordering drafts of verses from the regional
garrisons as well as from the private libraries of followers in Mecca
and Medina. He stipulated conditions for submissions to be accepted:
they must have first been written in the Prophet's presence, there must
be at least two witnesses to the Prophet's recitation and, where there
was disagreement over the wording within the editorial committee,
the dialect of the Quraysh, Muhammad's own clan, should be preferred. The tradition also affirms that this editorial committee ensured
that Uthman's text would be definitive, that all submissions would be
heard, that doubts over phraseology would be answered only by those
who heard the *ayas* (verses) directly from the Prophet and that the
caliph himself would supervise the work.

Uthman's legacy was therefore to create a unifying text to hold the
empire together, a book that acted as the religious authority in
Muhammad's stead. By the early 650s, the tradition recounts, Uthman had gathered the 77,000-plus words of the Koran, and scribes
spent four months then copying out the Uthmanic prototype. The
parent book was kept by Uthman in Medina, but four copies (their
parchment format now fairly well standardized across the empire)
were supposedly sent in the four directions of the compass, or, according to Islamic tradition, to Damascus, Basra and Kufa (the fourth city
would have been Eustat), with one copy kept at Medina. Uthman
then ordered all other copies of the Koran to be burnt.

Parchment formed the pages of Korans across the empire. It was
flexible and durable, making it an apt partner for the words of God,

since flexibility and durability both encouraged regular reading. When the parchment pages of a Koran were of the same size, it was called a *mushaf*; hundreds of *mushafs* survive today. The Koran itself makes indirect reference to two writing surfaces: parchment and papyrus. Papyrus, unlike parchment, was an import: only after the Arab conquest of Egypt, in 640, did the Arabs begin to use papyrus too. The oldest surviving Arabic text on papyrus dates to 642. The great fourteenth-century Muslim social historian Ibn Khaldun wrote that, in Islam's early days, parchments had been used for scholarly works, letters of government, bills of credit and other official notes, since people lived in luxury. Yet as a culture of books grew, he wrote, there had simply not been enough parchment available.

By the early ninth century, paper was increasingly playing a major role in the Caliphate. Korans had remained on parchment, perhaps because other writings needed to be portable whereas Korans simply needed to last. Paper, however, combined portability with durability. Moreover, a parchment Koran might need around 300 sheepskins, making it an expensive product. Paper was much more affordable. The earliest complete Arabic paper manuscripts are from the turn of the ninth century. It was an apt moment for paper to make itself felt, as the Caliphate was on the brink of a literary outpouring, one which might otherwise have been stunted by the price of its surfaces. The high status of the Koran may be one reason that it was slower than other Arabic books to make the switch from parchment, the expensive surface of wisdom and revelation for centuries; paper was often viewed as pedestrian by comparison. But in the tenth century even Korans switched to paper.

The earliest paper Korans used a script called *nakshi*. The script's shape led to the re-adoption of the vertical codex; previous Arabic scripts had been better suited to horizontal codices. The exception to the Koran's tenth-century adoption of paper was in the Maghrib, where parchment continued to carry the words of the Koran until the fourteenth century.

The Koran's role in the story of paper is unique. It was the cornerstone of a new empire and civilization that spanned much of Eurasia, one which would allow paper to sound the death knell of parchment and papyrus across the Caliphate. Yet the Koran became as much a recited monologue as a written book, and the memory and gestures associated with recitation were as crucial to its identity as were pen and ink. Its calligraphy can be viewed as a written recitation just as its recitation can be viewed as acoustic calligraphy. In part, this comes from its own roots. But, as the Caliphate spread, the Koranic message needed to be delivered to distant lands without being changed in the process. The result was a work that falls between two media. Even today, that tension continues, and silent reading is still considered second best to recitation.

In the first centuries of Islam, an instructor would teach or recite the Koran orally and the students would then try to retain it by memory. They would not read the teacher's notes, and when he died those notes would sometimes be burnt. Islamic tradition avers that written transmission was far from trusted and that the favoured medium for the Koran was not the page but the human voice. And yet, even if its transmission process did favour the oral record, it was the Koran, first a work to be heard and recited, then a work which remained wedded to parchment decades after other texts had switched to paper, which would form the cornerstone of a paper civilization.

The Koran cannot simply be set alongside the Christian and Hebrew scriptures as just one of the major monotheistic texts that aided paper's Asian fortunes prior to the Renaissance. Instead, the Koran has a much more significant, and unusual, place in paper's story. Both the Hebrew and Christian scriptures had spent centuries on papyrus, parchment or vellum before the arrival of paper in Western Asia. But, as we shall see, Islam, after committing itself to the page, made the switch to paper far more quickly. Moreover, it was not Christianity that conquered so much of Asia in the paper age – it was Islam. This meant that paper would have as its ally the highest text of a civilization that ruled over half of Eurasia, a civilization that would seek not just conquests for the Caliphate but, in time, converts for its book-centred religion, too. Christianity would undergo its own conversion to paper, but that conversion would come centuries later and attach itself not just to a cultural transformation but also to a major – and internal – theological shift. In the meantime, parchment and vellum would suit Rome and Constantinople well.

If the Koran enjoyed a unique power in propelling paper's conquest of Asia prior to the Renaissance, it was thanks not only to its role as the primary text of the Islamic Caliphate, which at its height under the Abbasid dynasty (750–1258) had a population of close to 20 million. The theology of the Koran was crucial too, whereas the Tanakh and Bible are not traditionally viewed by Jews and Christians as having always existed. The Koran, on the other hand, came to be viewed by most Islamic theologians as eternal and uncreated in its own right. In the early days of Islam, the nature of the Koran was *the* great subject that Islamic theologians debated, much as the nature of Jesus had been *the* great subject for debate among Christian theologians in the early Church. God's primary word to humanity in Christianity had been the person of Jesus, God-made-man, whereas for Muslims God's word to humanity is the Koran itself. In Islamic civilization, following decades of controversy, it gradually became orthodoxy that the Koran itself was the eternal word of God.

Thus Muslim theologians came to believe that the Koran was as uncreated as God himself, a teaching that began its life in controversy (an important element in disagreements between the Mu'tazilites and Ash'arites) but which was gradually accepted over the course of the ninth century. The result of this doctrine was that the book's provenance could not be questioned. This created a paradox, for as the words of the Koran spilled onto the doors, walls, chests, carpets, buildings, bindings and furniture of the Muslim empire, a raft of sciences took shape: physical, linguistic, philosophical and theological, as the next chapter relates. And yet for all these subjects of study and criticism, the Koran itself could not be fully set in its historical amd intertextual context. The response of one legal school to any questioning of God's word became *bil kayfa* – 'without how'.

This doctrine of the Koran is also why, after thirteen centuries, there is no critical edition of it that cites its source manuscripts and analyses that manuscript history. The 1924 Egyptian edition so widely used today is based not on a text-critical reading but a canonical reading – community acceptance through Islamic history, not antiquity, was therefore paramount. There are around 4,000 pages of Koranic text surviving from the first two centuries of Islam, which in the early *hijazi* script is the equivalent (in terms of quantity) of around seven

Korans. Yet attempts to return to the earliest texts and so to relay the Koran's early manuscript history have foundered. In recent decades, the scale and logistics of the project have presented a particular challenge. Yet there is clearly hunger among many scholars (Islamic and non-Islamic) for a critical edition that comes as close as possible to the exact phrasing and sounds of the original – or at least earliest – Koran. One Koranic scholar has described it as 'the most cherished dream of everyone who works with the Qur'an'.[61]

The Koran's unimpeachable authority gave to paper an unparalleled role, since it no longer simply carried the words of God – instead, it carried the uncreated words of God. Not simply eternal truths, but an eternal book. Once established as a binding for society, such a doctrine is not easy to challenge or even to chip away at. In the story of paper, it is a doctrine that has helped one of history's greatest civilizations to persist and to place the written word at its heart, with global consequences.

That status has made it difficult to subject the original compilation of the Koran to widespread academic scrutiny, even when miracle finds of documents do occur, as in the 1970s when a hoard of manuscripts, including some 12,000 Koran fragments, was found by builders in the roof of an old mosque in Yemen. The manuscripts, Koranic texts dating from the seventh and eighth centuries, were then studied by two Classical Arabic philologists, Gerd Puin and Hans-Caspar von Bothmer, who found in the documents minor textual variations in the *rasm* (vocalization) never before recorded. Puin concluded that there had been changes in the oral tradition of reading the Koran, and that hundreds of variants had been invented by Muslim grammarians, philologists and exegetes three centuries after the death of Muhammad. That view remains controversial among Koranic scholars, but the Yemeni authorities withdrew the documents owing to the controversy that comments about them was sparking in the Islamic world. The fragments remain unpublished, in part because of concern over perceived anti-Muslim motivations. But both Puin and von Bothmer believed they constitute the earliest manuscripts of the Koran; indeed, von Bothmer concludes that a complete standard text of the Koran existed in the first Islamic century.

*

These discoveries are not merely important for the Koran but also for the spread of paper. They begin to answer how it was possible for a largely oral culture from Arabia to spread a book-based religion across half of Asia. They begin to answer whether the Koran really was its own invention, or whether it took the influence of more bookish cultures to bring it about. Arguments over the Koran's nature, as an oral recitation or a written book, as a created revelation or an eternal book, helped to foster debates that found their way to the page, since the page could transport those debates right across the empire. Even the problem of agreeing on a standardized text, and all the challenges around the Koran's compilation, gave rise to a culture of written scholarship and the study of language, as the next chapter recounts. All these factors had a role to play in making Muhammad's new religious community a way of life founded on a single book.

Those books were, of course, not yet on paper. By the fourth century, parchment and vellum were increasingly competing with papyrus across Western Asia, and with that came the replacement of the scroll with the codex. Until then, Egyptian papyrus had dominated writing surfaces, but papyrus disintegrates more quickly and is unsuited to folding into a codex. Thus parchment and vellum became the surface of Islam's holy books in its early years, after the first scraps on the bark of date palms, on animal skin or, sometimes, on papyrus or bone. Papyrus was usually preferred for legal and commercial documents and tended to be kept in scrolls, whereas the codex was more portable and manageable. Moreover, Coptic and Syriac Christians were already using it for their own holy book, which began as a collection of parchment sheets between two boards.

To make a codex, each double folio was folded into two leaves, which were assembled in gatherings, then sewn together and bound as quires into a book or codex. In Muslim books, the text is in a single-column block, like rows of worshippers in a mosque, whereas in traditional European bibles, each page carried two columns, like pews either side of a cathedral nave. Paper only took over Korans very gradually but, by the tenth century, the shift had taken place everywhere in the Muslim world except North Africa.

Korans were often split into thirty parts, one for reading each day of the month, and in the eighth century a system of diagonal strokes was introduced to the Arabic script to avoid confusion between letters. But the Uthmanic Koran could not be fundamentally altered. Scribes therefore used red or yellow inks for any additions, so as to distinguish them from the black-inked letters of God.

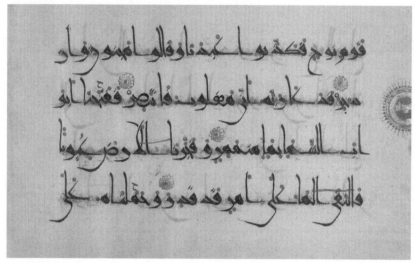

15 An early paper Koran. This non-illustrated manuscript dates to AD 993 (AH 383), was produced in Isfahan, Iran, and measures 9 by 13 inches. The text is written using black ink, the circular diacritical marks (indicating vowels) are in red, and gold is used for the decorative medallions. The script is a more angular version of Kufic than parchment Korans generally carried. It also points to new influences affecting the look of Islam's holy book – not the Arabian desert or the Eastern Mediterranean but the eastern lands of the Caliphate. (Copyright © Metropolitan Museum of Art/Art Resource/Photo SCALA, Florence, 2013.)

The history of the Koranic manuscripts is the history of the Arabic script. The script was derived in part from Syriac as well as Nabatean scripts: before Islam, Arabic script appears in only a handful of inscriptions. But from the seventh century it began to displace its competitors and became widespread enough to develop regional forms. Nevertheless, the bureaucratic elite worked hard to ensure linguistic unity as the glue of empire, and their Korans reflected this ambition.

Soon scribes sat at the centre of Islamic civilization. They would receive a diploma after several years' training under a master scribe. The master would teach them how to trim reeds for poems, how to prepare inks from carbon or iron gall or gum and how to illuminate their texts. Often the calligrapher and illuminator was the same person. At the least, the two formed a partnership in which they co-designed a page's layout. Al-Baihaqi, a tenth-century *hadith* expert from Khurasan, reported that those unable to write would turn up at the mosque, clutching sheets of parchment, in the hope that people would write for them. The *hadith* even say that, when the ink of the scholars comes to be weighed on the Day of Judgment, it will outweigh the blood of the martyrs.

Opening a Koran was like entering a sacred building. It began with a 'carpet page', often decorated with hexagrams and interlocking arabesques of text. After the conquest of Egypt in the 640s, books were carefully bound and often, thanks to Moroccan influence, tooled with gold. Inside, the handwriting unveiled the character of its scribe; the best handwriting was a signature of spirituality. On the pages themselves, gold was primary, but blues, reds, greens and yellows extended the decorations. Suns and trees were dotted across them: a tree representing the Koran stretched from earth to heaven, decorated with the blue and gold of the heavens and its lights.

In Egypt in the early fourteenth century, the first Mamluk dynasty Koran had ornate colophons for each volume.[62] The colophon for the seventh volume sat on a cloud motif against a pink foliate background, flanked with gold filigree on a blue setting. The patron, calligrapher and illuminator are all mentioned. Each volume is announced with a grand double frontispiece or carpet page. Its letters are entirely in gold. The book had become the glory of Islamic culture.

The illustrative and calligraphic beauty achieved in the Korans of the early and middle periods of Islamic history is mesmerizing. For Muslims, the Koran was God's book eternal, the word made parchment. When Caliph Uthman, the Koran's publisher-in-chief, was assassinated, blood from the sword wounds in his neck and head was said to have stained the pages of his first Koran, which he still held in his hands. But this sometimes fragmentary, sometimes inscrutable text, as important for its rhymes and cadences as for its content, for its role as a sacred object as for its status as a teacher, outgrew parchment soon enough. And, as it spawned the sciences of Islam's most productive era, it launched Asia's western half into the paper age.

12

Bagdatixon and its Ologies

My sons! Wherever you stand in the market before a shop,
stand only before those where weapons and books are sold.
                              Al-Muhallabi, tenth century[63]

As for the writing of books, no profession is more irksome;
Its shoots and its fruits spell penury.
He who undertakes it is like one who uses his needle
To clothe others while he himself remains naked.
                              Abdullah ibn Sarah, 1121[64]

Clues to the products, knowledge and genius of Islamic civilization lie scattered through English dictionaries in words we have borrowed from Arabic. Take, for example, your common citrus fruits: lemons, limes and oranges. Then there are the flowers, herbs, sweets and spices: sherbet, saffron, sugar, syrup, jasmine, lilac and marzipan. There are clothes and fabrics, from satin to sashes to cotton to cassocks to damask to muslin. There are animals, like the giraffe and the camel; drinks, like alcohol and coffee; and musical instruments, like the tambourine and the lute. There are foods and cooking techniques, like spinach and tandoori. And there is a ripe miscellany of many words besides, from azure to safari to amalgam, from hashish to Satan to mafia, from rackets to elixirs to crimson, and from mummy to magazine to checkmate.

This flurry of words arrived on European shores as imports. All of them came from the most scientifically and intellectually advanced civilization of its day, the Abbasid Caliphate, which ruled over the lands of Islam from its capital city, Baghdad, and had won battles as far east as Talas. Yet Abbasid greatness had little to do with muslin, spinach or safaris. For that, you must look to other terms, terms for which Europe had no words of its own, terms that form the titles of borrowed genius. In mathematics, these new arrivals included algebra, the azimuth, algorithms and zero. In the earth sciences, Arabic introduced alchemy, chemistry, aniline, carat and alkali to European tongues. In the study of the heavens, it parcelled out still more words, from astrolabe to nadir to zenith.

Not all these words, or at least not all the ideas behind them, were home-grown. Thus the Abbasids took algorithms and zero from the Hindus, just as they took alchemy and the astrolabe from the Greeks. But they did domesticate them, making them their own and, in many cases, ensuring their survival. They built vast libraries to store their knowledge, but they were more than mere curators. Instead, they read, debated, proposed, tested, analysed, explored and discovered too. The Abbasids fed on existing knowledge before reproducing, expanding and filing it. They even criticized the Byzantines for ignoring the wisdom of the Greeks and repeatedly asked them for Greek texts of natural philosophy to translate. It was not simply a translation from one language to another, however. The Arabic scribes copied the Greek manuscripts from papyrus and parchment onto paper.

The Abbasid heyday is one of history's golden ages, as the dynasty pursued philosophy, the study of the heavens (astrology, astronomy and cosmology), of language (poetry, linguistics and grammar), of the earth (chemistry, botany, geography and geology) and of mathematics (geometry, algebra and decimals). But more esoteric and reflective subjects caught widespread attention too, from magic and alchemy to theology, as well as more epicurean subjects, like cooking and erotica.

Classical Arabic is a difficult language to master. Grammar, lexicography, etymology and philology all developed under the Umayyads, Islam's short-lived first dynasty, and their successors, the Abbasids, as the outworking of Koranic studies. These linguistic sciences were the first stirrings of an age of inquiry led by writing, and the trigger for them all had been the Koran.

The explanation of the Koran, *tafsir*, was considered a science in its own right, as was the study of Islamic law. Moreover, some of the laws threw up scientific conundrums too. How could the muezzin be sure that he made his five daily calls to prayer at the correct times?

How could an architect check that the *qibla* (the prayer-niche of a mosque) pointed towards Mecca? Only engaged, dedicated science and mathematics could provide accurate answers to these questions right across the Caliphate.

Metropolitan muezzins learnt to tell the time by careful study of the stars, often using instruments to help them, and their handbooks listed prayer times in cities throughout the empire. Yet these forms of knowledge were more than merely reactive. Instead, sums to determine the *qibla*, the direction of Mecca, were computed by men already conversant with trigonometry, astronomy and geography. These findings then had to be written down. As philosophy, medicine, astrology and astronomy developed alongside the linguistic sciences, and as writing flourished, the limited supply of papyrus and the expense of parchment quickly became a problem.

The eleventh-century historian al-Tha'labi created an extensive inventory of the empire's market goods, matching products to their places of origin. In the Levant, he listed the cottons and papyrus of Egypt, the apples, glassware and olive oil of Syria, the swords and cloaks of Yemen, the robes of Rum and the roses of Jur. In the Caucasus and Persia, he praised the carpets of Armenia, the honey of Isfahan, the clothes of Merv and the cloaks of Rayy. Still further east, he named the musk of Tibet and, finally, the paper of Samarkand. Ibn al-Faqih, a tenth-century historian, wrote that the people of Khurasan were so expert in papermaking that their country might as well be a part of China itself. Yet papermaking would not remain confined to Khurasan and East Asia for long, thanks in particular to the Barmakid family.

The word Barmecide (from Barmakid) is another Arab import; in English, it describes something illusory or imaginary. It derives from a story from *The Arabian Nights*, in which a member of the Barmakid family offers a meal of many courses to a beggar, a meal in which each dish is in fact empty. The beggar, Shacabac, is able to see the funny side and so pretends to eat the dishes. This was the first Barmecide feast.

Other than the Abbasids themselves, the Barmakids were Baghdad's most prosperous and famous family and, perhaps unsurprisingly,

they were from the east. For generations, their family had been guardians of the great Buddhist shrine at Balkh in northern Afghanistan,
one of the major metropolises of the Middle Ages. The shrine was a
magnet for Buddhist pilgrims and among the religion's holiest sites.

Around the 660s, the Barmakids had married into the royalty of
Transoxiana to the north and converted to Islam. The conversion
proved political too, as the family switched its allegiance to the
Abbasids. When the Abbasids took power in 750, Khalid ibn Barmak
threw in his lot with the bureaucracy and never returned to the east.
Instead, he became the financial administrator of the early Abbasid
Caliphate and was appointed to the governorship of Fars in Persia,
where he quickly became popular and even turned his hand to archae-
ology, storing ancient Persian treasures in a mountaintop refuge that
the Arab armies would not attack. Back in Baghdad, he persuaded the
caliph not to destroy the great arch of Ctesiphon, erected by the Persian king Chosroes II in the seventh century. This rediscovery of
Sassanid Persia was one of the driving forces of both scholarship and
cosmopolitanism. There was even an Iranophile movement, a reaction
against the dominance of the Arabs. The tolerance and inclusiveness
of the Barmakids may have been symptoms of their Iranian loyalties
or even of their private coolness towards Islam (although if such coolness existed they were careful not to let it show). In their home,
theologians and free thinkers would meet to debate theology, linguistics and the sciences. Khalid also introduced the codex to replace the
scroll in the taxation and military bureaus.

When Khalid ibn Barmak died in 780, his sons were already senior
civil servants. (His son Yahya was so impressive that the caliph
quipped that, whereas some men beget sons, Yahya had begotten a
father.) Yahya ibn Khalid was appointed grand vizier in Baghdad
under Caliph Harun al-Rashid, the picaresque hero of *The Arabian
Nights*. His own sons, Fadl and Jafar, rose to senior positions too,
until the Abbasid administration had become a Barmakid affair. The
caliph even handed his personal affairs to Yahya, and the Barmakids
skilfully administered the empire and encouraged the arts, allowing
the caliph to play a largely ceremonial role.

It was Yahya's two sons who brought paper to the heart of the
Abbasid empire. Papyrus had quickly partnered with Islam and with

Arabic but, aside from the island of Sicily, quirks of soil and climate had granted Egypt the monopoly on production. Yet it could not meet the demand for writing surfaces across the empire. In the 830s the caliph had even tried to found a papyrus factory a little north of Baghdad but it had not been a success. Besides, papyrus was good for scrolls but ill-suited to codices, as its edges frayed too easily, and Islamic literature tended to sit between two covers, a habit it had acquired from Syriac Christianity.

Parchment, however, could work only so long as scribes and authors were few in number, and given the outpouring of bureaucracy and learning even in its first half-century, the Abbasid Caliphate could not find all the parchment it needed, nor could it afford it. There was another problem: writing was the medium for Abbasid governance across the empire, but, with the deft use of water and a cloth, writing on parchment could be wiped off the page without a trace. And if words could disappear, they could also be replaced.

The two Barmakid brothers decided that the Abbasids needed to make the switch to paper. Fadl was governor of Khurasan, home to the city of Samarkand and its famous paper mills. The switch was surely his idea, but it was Jafar, the grand vizier, who took the decision to convert the empire to the paper of Khurasan. Jafar knew that the catalyst of such a transformation could never be the scholars, the booksellers or even the theologians, but only those High Priests of the Pax Islamica: the Abbasid bureaucrats.

Ibn Khaldun, the great North African historian of Islam, wrote in the late fourteenth century that parchment had no longer been sufficient. The reason, he argued, had been the expansion of both bureaucracy and scholarship.

> Paper was used for government documents and certificates. Later, people used sheets of paper for state and scholarly writings, and its manufacture rose to new heights in quality.

It was still possible to import high-quality Chinese paper, but it was too soft, designed for the brushes of East Asia not the pens of the Muslim Caliphate. Moreover, Silk Road goods were expensive. Centuries earlier, Pliny the Elder had written that goods increased in value by a hundred times as they passed along the Silk Road from east to west and, as if to prove him right, the eleventh-century Muslim calligrapher Ibn al-Bawwab once accepted a stock of Chinese paper (presumably no more than 200 or 300 sheets) instead of a hundred gold dinars and a robe of honour.

Importing paper 1,300 miles from Samarkand was hardly a long-term solution. Fortunately, Baghdad was itself the greatest metropolitan invention of the eighth century and it was ready to learn everything it could. It radiated from the Round City, a geometrically planned symbol of power, order and learning which its founder, al-Mansur, built as a statement of intent. He wanted Baghdad to become the greatest city on earth, the world's intellectual and scientific superpower – and he achieved his ambition. The first Arab institute of science was built here, while books were gathered in from neighbouring kingdoms and the city became a hub for scholars and scientists from distant lands.

In 795 Baghdad finally built its own paper mill. After centuries marooned in East Asia, papermaking had spread from China's western borders to Mesopotamia, some 1,400 miles in less than fifty years. Moreover, the Baghdad mill produced enough paper for it to replace papyrus and parchment as the surface of the civil service's bureaucratic records.

The Barmakids barely lived to see the beginning of the knowledge revolution they had wrought. There is still uncertainty over what led to their fall from favour but it may have been the discovery that Jafar was having an affair with the caliph's sister. Whatever the cause, in 803 the caliph ordered one of his pages, Salam al-Abrash, to confiscate Jafar's goods. When Salam arrived at Jafar's house, the curtains had already been pulled down and Jafar himself was complaining that it was as though the apocalypse had begun. In the end, he was beheaded. The fall of the Barmakids robbed Baghdad of its most colourful and cultured family and the remaining six years of Harun's reign were lacklustre by comparison. But one of their greatest legacies was the adoption of papermaking – there was even a type of paper called *Jafari* – and the city's paper soon became known elsewhere too. Some Byzantine writers named it *Bagdatixon*.

Papers were exported outside the Muslim empire and even into Europe. Paper attracted writings on subjects in those foreign lands too. One of the earliest surviving writings on Arabic paper is a Greek manuscript on the teachings of the Church Fathers called *Doctrina Patrum*. It comes from Damascus and dates to around 800. The oldest extant and complete Arabic paper book dates to 848 and was found at Alexandria in Egypt.

Islamic papers were usually made from a recipe of linen and hemp, which made them strong, durable and opaque. (The earliest Islamic papers were thick and weighty too; later on, a better mashing process improved their quality.) Rags and cords were unravelled and softened by combing before being steeped in lime water and then hand-kneaded to a pulp and bleached. The pulp then progressed from the mould onto a smooth wall to drain it, until it fell off. Next it was rubbed smooth with a starch mix and later dipped in rice water to clog the pores, thus holding the fibres together more firmly. Sometimes just one side was smoothed for writing so that two sheets could be stuck to one another for added strength. The paper came in three grades: common, raw or straw. It could also be given a range of finishes, from glazed to glossy to smooth.

When a buyer received his order, his sheets had been pre-folded to the size requested. They arrived in packs of twenty-five called *dast*, the Persian word for 'hand', which was translated into *kaff* in Arabic and, later, into *main de papier* in French. Five of these hands formed a *rizma*, from which comes the word 'ream'.

After choosing the appropriate quality of paper, a calligrapher would combine it with *ahar*, a mixture of rice powder, starch, quince kernels, egg white and other ingredients, to give the paper a shiny surface which the pen could glide over easily. The calligrapher then burnished and smoothed the paper with a stone and placed a *mastar* between two sheets. (A *mastar* was a series of silken threads set across a cardboard frame to act as a ruler.) When he finished his writing, he sprinkled some sand over it as a ritual blessing.

Choice of colour was especially significant. Blue paper was used for mourning and even for death sentences in Egypt and Syria, whereas red was preferred for festivals and deep red for letters between senior officials. Baghdad even produced its own characteristic size, known as *Baghdadi*, which measured 42 by 29 inches, but the range of sizes trickled right down to the 'bird paper' that was attached to the fixed feathers of carrier pigeons: these papers measured just 3½ by 2½ inches. (It was also paper that allowed the shift to smaller, more personalized Korans, even pocket-sized Korans, always a turning point in the culture of any book-based religion.) If the paper looked antiquated, which some buyers preferred, it was because it had been treated with saffron or fig juice.

Paper raised the profile of the bureaucratic scribal class in general, known as the 'people of the pen', who kept their writing sets in decorated boxes which they carried on their waistbands. But it raised the profile of calligraphers in particular, and the Islamic injunction against representing God in an image (combined with Islamic coolness towards figurative art more generally) made calligraphy a more appealing art form. The greatest Abbasid calligrapher was Ibn Muqla, who was vizier to three caliphs in the tenth century and invented six script styles for Arabic. But his involvement in politics led to his lower right arm being amputated. It was said that he still managed to attach a reed pen to what remained of his arm and to compose calligraphy as beautifully as ever.

A trainee calligrapher needed to study under a calligraphy master for months or even years before he received the certificate that allowed him to sign products with his own name. The master taught him how to sit, usually squatting but sometimes kneeling with his feet folded backwards under him. He taught the student to rest the paper on his left hand or on the knee so that it was slightly flexible, as the round endings could be written more easily in this way than if it was placed on a hard desk or low table.

The student had to learn to curve some of his letter-endings until they appeared 'woven on the same loom'. He was taught to trim his reed pen, although he might also use a fine steel pen or a quill. If he received the certificate, he could be sure of respect, but he would have to be exceptional to secure a place in a royal library, which would earn him a title: 'Model of Scribes' or 'Golden Pen'. The first great Abbasid calligrapher was known as 'the Squint-Eyed'. Perhaps the long hours spent pen-in-hand had spoiled his eyesight. In the fifteenth century, the great Persian calligrapher Mir Ali recalled forty years he had spent in calligraphy but complained that, while it had been slow to learn, yet without practice the skill was quickly forgotten.

This new passion for writing began, however, not with single-minded calligraphers but with bureaucrats. The Abbasid state used writing as its means of political control across the empire and its secretarial corps needed to be familiar with good grammar as well as with the administrative literature of bureaucracy, from technical terms for pens and inkwells to how to seal a document or keep a tax register. They used these skills to keep records of income and expenditure as well as military pay levels. Such men were trained in stylistics and orthography so that they could use and preserve the best Arabic possible as glue for the empire.

In fact, the Abbasids developed an abstruse bureaucracy so loaded with formalities and protocol that only experts could compose official documents of state. Moreover, in the provinces, registers had to be kept in different languages: Pahlavi in Mesopotamia and Persia, Greek and Syriac in Syria, Greek and Coptic in Egypt. Scribes quickly rose in stature under the Abbasids and it was only these same scribes who could have invented the new 'stationer's script', an angular script coined for copying onto paper – the product of people who knew the surface well. By the mid-tenth century a few of them were even writing books about the bureaucracy itself – its development, achievements and even its heroes – or simply books of advice to their fellow scribes, like *The Craft of Writing* and *The Education of the State Secretary*.

Their aim was hardly original, namely, to govern the state by writing. But this did expand the use of paper, whether for tax records, legal documents, official correspondence, state archives, postal missives or military registers. Paper became the field of engagement for taxes, army services and the myriad new offices in Baghdad: the War Office, the Office of Expenditure, the State Treasury, the Board of Comparison, the Office of Correspondence, the Post Office, the Cabinet, the Office of Signet, the Office of Letter Opening (the caliph's inbox), the Caliph's Bank and the Office of Charity.

Fortunately, there was plenty of linen and hemp to feed this new culture of paperwork and books. The population was probably able to offer a continuous supply of rags to recycle, a supply that fuelled the mills that sold to the bookmakers, bookshops and libraries throughout the city. Walking through eleventh-century Baghdad, you had a choice of more than a hundred bookshops.

Most of the bookshops lay in the south-west of the city, in the *Suq al warraqin*, the Stationers' Market, feeding and expanding the demand for paper. There were paper shops here too, as stationery was now one of the city's great trades. (Some of the mills on the River Tigris may have been paper mills.) By the mid-ninth century, many educated Muslims, Christians and Jews in the Abbasid Caliphate were using paper to write letters, to keep records and to copy out literary or theological works.

Book collections began to sprout across Baghdad, as the readers' market grew. Some 600,000 hand-copied Arabic manuscript books survive from the pre-printing era and this number can only be a small proportion of the number actually produced. There were public libraries and fee-charging reading rooms, since paper had made books cheaper but not yet cheap enough for all to afford. Calligraphers and copyists were the beneficiaries.

Bureaucrats are not usually dubbed pioneers, but in the Abbasid Caliphate it was bureaucrats who drove a new, literary culture until good diction and good calligraphy became marks of good breeding. Mudari Arabic, as the post-classical language is known, gathered to itself a galaxy of words and wordplays for men of letters to draw on. In a few cases, women took up the pen too, especially among those women based at court. At the roots of this new passion for the Arabic language lay not just the Koran itself, reckoned the highest authority on good Arabic, but also the scholars who studied Arabic grammar and vocabulary for more technical purposes. The tenth-century philologist al-Bawardi, who died in 957, dictated 30,000 pages on linguistic topics from memory. From such scholarly beginnings, more frivolous uses of language began to emerge too.

Men were admired for wording their thoughts playfully and intricately while their letters of state were expected to be lessons in elegance, feathered with rhyming prose. Any serious composition included quotations from poetry and references to obscure points of learning. Often men wrote about, say, biology, not to explore and explain the animal world but to amuse their readers with well-worked phrases and witty observations. At court, writing was still more frivolous and poems even composed about the food at state banquets.

Meanwhile, the Arabic language was taking over from Greek as the Mediterranean's storehouse of ideas and science. As the translations into Arabic poured in – of Greek and Persian philosophy, Indian mathematics, Jewish and Christian scriptures and works in Pahlavi, Greek, Sanskrit, Hebrew and Syriac – the Abbasid bookshelves became a library of Eurasian learning and ideas.

Caliph al-Mansur, Baghdad's eighth-century founder, devoted himself to literature, opening a translation bureau and collecting Greek, Persian and Sanskrit works in their hundreds on philosophy, medicine and astronomy, to name just a few of their subjects. All were translated into Arabic. Bibliophiles built private libraries. A popular explanation of the death of the ninth-century scholar al-Jahiz, already partially paralysed in his old age, was that he had piled so many books around his house that one day they collapsed and killed him. Another Baghdad bookworm had his sleeves enlarged so he could carry bigger volumes in them as he walked through the city.

Baghdad had no professional academic class, as we would understand it, but the city was full of unofficial intellectuals, of people absorbed in books, copying, book dealing, reading and in collecting the Traditions of the Prophet, the *hadith*. The influence of the Koran was far-reaching: Koranic injunctions to heal the sick propelled medicine and led to free local health care, as well as to new drugs and to advances in optics and surgery.

The Caliphate's first Institute of Scientific Studies (mentioned earlier) was founded in Baghdad in 830, a fusion of a library, an academy and a translation bureau. It was known as the House of Wisdom. After a dream in which he claimed Aristotle had appeared to him, Caliph al-Mamun, who governed the empire from 813 to 833, sent letters to the Byzantine emperor asking for the works of Aristotle, Plato, Galen, Hippocrates and Ptolemy. Once he received them, he commissioned the most experienced of his translators, notably al-Kindi (author of more than 260 titles, according to Ibn al-Nadim) to produce Arabic versions. He also sent astronomers to work in the House of Wisdom, its library inherited (if only in its conception) from Sassanid Persia and home to many translated scientific works.

Al-Mamun collected rare books from anywhere he could, sending scholars to Egypt, Syria, Persia and India. Some went of their own accord, as the empire allowed them to travel and scholars were increasingly curious to learn from other cultures. One bibliophile, Husain bin Ishaq, travelled to Palestine, Egypt and Syria in search of a single book – in the end he found half of it in Damascus. But foreign scholars came to Baghdad too; the Hindu physician Duban worked for al-Mamun alongside Parsees, Christians, Jews and Muslims.

Caliph Mamun's scholars calculated the earth's circumference with remarkable accuracy. The West would name algorithms after the philosopher and mathematician al-Khwarizmi. He also resurrected algebra and introduced the numerals we use today, which led to the discovery of decimal fractions and the calculation of pi to six decimal places. Five of the six trigonometric functions (cosine, tangent, cotangent, secant and cosecant) were Arab discoveries, engineered on the basis of the Hindu knowledge of sine; these were the building blocks for mathematical astronomy.

The Arabic translation of the second-century *Almagest* of Ptolemy was seminal for astronomy. Mapmaking and navigation progressed apace, proving that the Indian Ocean was not landlocked and thus indirectly fuelling Europe's Age of Exploration. Astronomical data, which were still read in Roman fractions in Europe, were calculated far more accurately by the Arabs in degrees, minutes and seconds. Abbasid science was fostered and encouraged by the state. The Arab search for the laws that governed nature was also a religious quest, stemming from belief in a divinely appointed order that could be studied, understood and harnessed to the religious needs of the state – from helping pilgrims with directions on the *hajj* to the tenth-century scholar Abu ibn al-Anbari's fabulous 45,000 pages of *hadiths*.

Improved access to paper as a writing and reading surface aided standardization, spurring improved systems of notation in mathematics, geography and genealogy. In this growing complexity, it advanced the capacity of written communication too, aiding clarity. Paper even transformed metalwork, ceramics, textiles, pottery, weaving and architecture, since it allowed designs to be drawn out first as well as for them to be sent on to fellow artisans thousands of miles away. (It has been argued that Persian miniatures, oriental rugs and the Taj Mahal could probably never have existed without this cultural exchange that paper allowed.[65]) In short, paper helped to create an empire-wide network of knowledge and learning that became the engine-room of Islamic civilization, and its availability cemented Arab Islam's adoption of the page as its storehouse of knowledge, where in the past it had simply relied on oral tradition. If the late eighth century saw the rise of paper across the Caliphate, it was the ninth century that saw its potential increasingly realized in Islamic civilization's great crescendo of learning.

To the north, the Byzantine empire only adopted paper after the Caliphate: paper may have seen some use in the ninth century but it was certainly not widespread before the eleventh and the re-use of parchment in palimpsests continued, even in Constantinople. The Byzantines viewed themselves as the preservers of ancient Greece's literary heritage but they were only partly successful; there is no evidence their libraries ranked with the largest of the ancient world. Instead, several hundred books were viewed as ample for a major Byzantine library, while the great Abbasid libraries' stock could be counted in the thousands or tens of thousands. The Byzantines tended to import their paper, initially from the Arabs, later from Spain, and still later from Italy, and even after the eleventh century, paper was used in state archives and not for religious manuscripts. Paper slowly became more common in Byzantium during the thirteenth century, but only came to dominate Byzantine writing in the fourteenth. By then, book collections had been severely reduced as a result of the sacking of Constantinople by the Crusaders in 1204. By 1453, when the Turks conquered the city, there were very few texts left.[66] It was only at this point that a paper mill was established (by the Ottomans) in Constantinople itself.

To the south and east of Byzantium, however, papermaking developed a far more impressive momentum, one that would ultimately drive it into Europe. From Baghdad, it was able to travel through Egypt and the Maghrib with ease. Indeed, southern Europe would show a strong interest in Islamic paper before the Byzantines, and Syrian paper became famous in Europe as *charta damascena*.

Constantinople was, at least in terms of the Mediterranean, the end of one of paper's many trails.

While European libraries still counted their books in the hundreds for most of the mid and late Middle Ages – even as late as the fourteenth century the Vatican Library held only 2,000 volumes (largely on parchment or vellum) – Islamic lands built libraries in all their major cities. In Baghdad, a theological school founded in 1065 cost 60,000 dinars to build but had an annual expenditure of 60–70 million dinars. Its books numbered in the tens of thousands. In 1228 a new institute was built in eastern Baghdad and 160 camels were needed to transfer its rare and valuable books from the Imperial Library. It used an open-shelf system and students could even access rare manuscripts. It had taken six years to build and it included lecture rooms for teaching astronomy and other sciences, as well as the teachings of Muhammad. At its peak, it held 140,000 volumes.

In North Africa, a royal library was founded in Cairo late in the tenth century. By the twelfth century, under the Fatimid dynasty, it was counted as one of the wonders of the world. It had forty rooms and 1,600,000 books and booklets; 600,000 of these dealt with theology, grammar, tradition, history, geography, astronomy and chemistry. There were 12 copies of al-Tabari's seminal history and 2,000 copies of the Koran written by famous calligraphers. Meanwhile, also in Cairo, the Al-Azhar Mosque Library held a stock of some 200,000 books in the late tenth century. A record of the royal library's annual budget breakdown has survived, proof that paper, though far less expensive than parchment or papyrus, was still not cheap:

> 275 dinars
> Librarian salary 48 dinars
> Paper for copyists 90 dinars
> Paper, ink, pens 12 dinars
> Torn/damaged book repair 12 dinars[67]

In 1068, as political and economic instability plagued the Caliphate (following several decades during which the Abbasids' control was on the wane), twenty-five camel-loads of its books were sold for just 100,000 dinars to pay soldiers' wages. A few months later Turkic soldiers plundered and destroyed the rest, burning some and throwing others into the Nile (although some manuscripts would be rescued). They tore up the leather bindings to make shoes and discarded much of what was left at a site they later dubbed 'Book Hill' – biblioclasm, usually in the form of bookburning, is as old as the book itself.

Libraries and paper mills often grew up together and the chief Islamic papermaking centres were often the heartlands of its literary culture and book trade too. To the east, of course, there were flourishing Islamic papermaking industries in Samarkand, Tabriz in Persia and Daulatabad, in north-west India. At the southern end of the Arabian peninsula, paper was made at Sana and Tihamah, both in Yemen. To the west, it was made in Xativa (now San Felipe) in Spain, Tripoli and in Fez, where one Arab historian counted 472 paper mills in the twelfth century. To the north, papermaking centred on Tiberias in Turkey and Hierapolis in Israel, as well as on the inevitable centres of Damascus, Cairo and Baghdad.

The tenth-century Abbasid emir Adud al-Dawla built a library in Shiraz, in southern Iran, with one long, vaulted room flanked by store rooms and high scaffolds covered in shelves, each scaffold devoted to a different subject. He equipped it with catalogues for reference and a ventilation chamber to carry water around its pipes. This last invention was less impressive, as the damp seems to have done for the collection. Just to the west, at Basra in today's southern Iraq, the city's library held 15,000 bound volumes (plus unbound books and loose manuscripts). Rayy, Mosul and Mashhad held large collections too.

The Caliphate had become a bureaucratic behemoth, a repository of ancient knowledge, a forum for authorship and a patron of philosophers and scientists alike, and paper had become the currency of information and communication from India to the Maghrib.

The Mongol invasions of the thirteenth century forged fresh lines of communication and commerce across Eurasia. Moreover, the curiosity of a handful of pioneering and radical European scholars had already led them to study the books of the Arabs, and even to travel to the Near East. As the twelfth century gave way to the thirteenth, this slow stream of scholars increased. They brought back Arabic books to translate into Latin, among them the works of Averroes and

Avicenna, whose natural philosophies would begin to unravel the Roman Church's intellectual authority, while learning was gradually shifting from monasteries to universities. In *The House of Wisdom*, Jonathan Lyons writes that Europe even imported an 'Arab Aristotle', an Aristotle tailored to suit the needs of the monotheism shared by Jews, Christians and Muslims. In other words, alterations made by the Abbasids could later become part of accepted European knowledge.

Hints of where paper was headed next in its worldwide journey could be found in Muslim Spain, where eight centuries of Islamic rule and frequent prosperity brought untold treasures of knowledge to the European mainland. Iberia had been a backwater before the main Arab invasion of 755, but it soon grew independent of the Abbasid cultural mainstream (as well as politically independent) and even began to compete for literary and scientific prowess. In the tenth century, Hakam II's library in Cordoba reportedly held 400,000 volumes. Even the library's index of authors and titles ran to forty-four volumes of fifty folios each. In Cordoba in southern Spain, private libraries flourished, as did non-Muslim libraries – some Christians kept libraries but most of their books were in Arabic. The city of Cordoba had a lively book market and, unusually for its time, female scholars.

Spain's incorporation into the Pax Islamica made it the recipient of waves of people, arts, plants, inventions, recipes, ideas and foods from across the Caliphate. Political division in early ninth-century Baghdad also brought a handful of fresh scholars to its shores. By the eleventh century, the Spanish had become Europe's best farmers and had begun to study Aristotle in detail. Thus Spain under Hakam II was as prodigious in learning as in libraries. Moreover, from Spain, as from the heartlands of the Caliphate, translations, commentaries and scientific treatises began to seep into some of Europe's first universities, in Bologna, Paris and Oxford.

Although Islamic Spain had begun to develop quite distinctly from the Abbasid Caliphate as early as the 750s, yet for the story of paper, its embrace of knowledge and learning was part of the same story that had begun in Baghdad. Despite its own paper spring, it would not be Islamic Spain which would transform Europe with its exported papers

– that transformation would come much later than the start of Al-Andalus's slow decline, which began in the eleventh century. Baghdad, however, remained the symbol of all that the Abbasids had achieved, a city which had gathered learning from around Europe and Asia – and put it to use. By the time the Mongol armies reached the gates of the old Abbasid capital in 1258, Baghdad boasted thirteen major libraries, including a madrasa library, founded in 1233, to which the caliph's porters had carried 80,000 works.

For a week the Mongol troops burnt the city's buildings, raped its women and ransacked its libraries. They threw Baghdad's books into the River Tigris and the waters were said to run black with ink for six months. Baghdad would rise again and even relaunch its papermaking industry, but, just as its days at the pinnacle of a united Caliphate were gone, so its role as the engine-room of paper's spread was finished too.

The ink and papers of its stored knowledge, science and ideas were swept downstream towards the Persian Gulf. Yet it would have been more apt had they flowed westwards instead, across Jordan and Palestine and into the Mediterranean Sea. For, after centuries in the Caliphate's intellectual shadow, Europe was drawing closer to a great paper-made revolution of its own, one that sat solidly on the shoulders of what the Abbasids had achieved, yet which would soon far surpass it.

13

A Continent Divides

*[He] ... is not simply one publicist within a larger constella-
tion. Rather, he was the dominant publicist. And he dominated
to such a degree that no other person to my knowledge has
ever dominated a propaganda war and mass movement since.
Not Lenin, not Mao Tse-Tung, not Thomas Jefferson, John
Adams or Patrick Henry.*

Mark Edwards, Printing, Propaganda and Martin Luther[68]

In the summer of 1521 *Junker* (Knight) George, a stout man in his
early forties, sat alone in a small room in Wartburg Castle, writing.
Perched on a cliff face 1,300 feet above the surrounding countryside,
Wartburg Castle lay not far from Eisenach, in the German heartlands.
George's room overlooked the forested Thuringian hills, and he would
often see crows gathered in their oaks and birches. The castle itself,
which still stands today, follows the spur of the ridge and is shaped
like a whale, its earliest part built in the twelfth century. It is a magnificent, lonely setting.

George was suffering from insomnia, depression, constipation and
sexual frustration. When he did manage to sleep he would have dark
dreams; by day, he was lonely and missed his home town. He was in
hiding from the authorities yet it may have been his local ruler who
had protectively orchestrated his 'kidnap' from imperial guards. This
same prince had sent him to Wartburg Castle to lie low for as long as
possible until the dust had settled. Stowed away in the old fort, George
lost no time in self-pity. Instead, he simply wrote.

He had brought only a few books and he wrote to friends that he was idle. But it was a strange idleness. He wrote sermons for every Sunday of the year, treatises on myriad areas of Church life, tracts on celibacy and monasticism and theological arguments addressed to the University of Paris. He wrote on German and European politics too. He wrote letters of advice to friends and old colleagues. He even began to translate Greek and Middle Eastern texts. In just ten months in the castle, he produced three volumes' worth of written papers. His sea of writings, which ran to sixty-seven volumes over the course of his lifetime, were part and parcel of Europe's new paper age. His real name, of course, was Martin Luther.

Prior to the paper age, George would have probably needed to find several libraries just to access even a handful of the books he needed. Many of Europe's medieval lending libraries were so ill-equipped to supply readers that they used a *pecia* system, whereby books were split up, usually into four quires, so that students could simply borrow one quire at a time, thereby speeding up copying. At least eleven major European university libraries used this system to improve access to books, from Oxford to Florence; the *pecia* system was the result of a new type of reference reading becoming popular across Europe – today we use the word 'research'. This moved book production out of the scriptorium as scribes could work on different parts of the texts in different places. It also raised the profile of the lay stationer, and coincided with the growing clout of the universities in pronouncing on theological matters, most notably the Sorbonne in Paris. Libraries were modernized still further when free monastic labour (in which the only payment was the promised remission of sins) was replaced with wage labour, a crucial move towards market mechanisms.

There had been significant libraries in Europe for centuries, thanks in large part to the Carolingian Renaissance in the ninth century, a movement that owed some of its momentum to Charlemagne's sponsorship of scholarship and increased education towards the end of the eighth century. In eastern France, western Germany and northern Italy, old Latin texts had been studied and a project started to teach Latin to the Anglo-Saxons. The greatest Carolingian libraries, in monasteries, cathedrals or royal courts, owned some 500 volumes each. Without the Carolingian recovery of classical texts, which sought to preserve the Roman classics, the only Latin writers we could access today would be Virgil, Terence and Livy.[69] Great medieval universities like Bologna and Padua also contributed in large measure to the regular reproduction of books. It was Walafrid Strabo, the ninth-century theologian and writer, who first divided books into chapters. The Carolingians laid the foundations for indexing, the interest in early classical manuscripts and the mass production of books: 7,200 Carolingian manuscripts survive from the ninth century.

Yet even the Carolingians were limited by their writing surface. Some wrote on wax tablets, as you could delete the text by simply melting the wax – the origin of the phrase 'a clean slate'. But you had to lift the stylus off the wax every time you needed to change its direction (even within a single letter) and the size of the wax tablet meant it was practical only for note-taking. For anything longer, medieval European scribes wrote on parchment or vellum, usually made from calf, goat or lambskin. Vellum was durable but it was also expensive and hard to treat. So long as books remained luxury items kept in the hands of a small and wealthy elite, vellum and parchment could be their fodder. (There was also some limited writing on papyrus in Europe, dating back as far as the eighth century, although the only producers were Egypt and Sicily.) But by the fourteenth century the exclusivity of books and learning was already unravelling in Europe. Vellum and parchment were soon to become anachronisms.

European papermaking began in Muslim-ruled Spain, where a stamping-mill was set up by 1056 in the Spanish town of Xativa (San Felipe), to macerate rags. Papermaking was an Arab import and water-driven trip-hammer mills were probably introduced to Iberia by the Arabs too,[70] who had used them for centuries. Manuscripts found at the monastery of Santo Domingo de Silos in the north of the peninsula prove that Spain had been using paper since the tenth century (presumably importing it from the Middle East). When domestic papermaking began, the process remained largely unchanged from its origins. Spanish papers tended to carry zigzag marks (from the draining gauze) and to be glazed, though in these details they merely reflected their Arab manufacturers' preferences. Pages were often thicker in the centre, as if the mould sagged in the middle. One manufacturing difference from East Asia was that, where bamboo and grasses had once been used for the moulds, European papermakers used metal gauzes, which left more pronounced laid and chain lines on the paper. (Laid and chain lines are the lines formed across and along the page by the grille it dries on.) The Xativa paper mill was soon followed by another in Toledo, and more in Catalonia and Bilbao (and by the first incontrovertible example of a water-powered mill, also in Xativa, in 1282).[71] Spanish papers were exported around the Mediterranean, to Morocco, Italy, Egypt and Byzantium.

It was also in the 1150s that paper was imported from the Middle East to southern Italy, but it was not widely used until the 1220s, when Germany began to import it too. In 1231 the Holy Roman Emperor, Frederick II, ruler of Italy, Germany, Burgundy, Naples and Sicily, banned the use of paper for all public notices and records in Naples, Sorrento and Amalfi, considering it less durable than vellum and parchment. But within government circles there was already a movement among scriveners to use paper rather than parchment.

There was some small-scale papermaking under way in northern Italy by 1235, but it was not until 1276 that the first significant Italian paper mills were set up in Fabriano, which sits close to the Adriatic Sea in northern Italy. They were followed by a mill in Bologna, established in 1293, where sheets of paper were six times cheaper than sheets of parchment. The Fabriano mills acted as a launch pad for European papermaking: by the 1350s the city was famous for its paper, which it traded across the Adriatic to the Balkans as well as to southern Italy and Sicily. In fact, many of the Fabriano mills were simply converted from older grain mills. Moreover, European millers were especially efficient in using water power thanks to their use of overshot wheels, where the water flows into the apex of the mill wheel, dropping several feet down and thus adding the force of gravity to the power of the current. (Fabriano was also where gelatine-sizing was first used, giving papers a tough finish well suited to Europe's quill-pens.)

By the 1340s paper was being made in the Saint-Julien region of France. In 1390, Ulman Stromer set up Germany's first paper mill in Nuremberg, dramatically reducing Germany's reliance on imported Italian papers. More mills followed around the turn of the fifteenth century in Ravensburg and Chemnitz, in Basel and Strasbourg in the mid-fifteenth century and in Austria, Brabant and Flanders towards the century's end. Papermaking began in Poland and England late in the fifteenth century. By this time, both countries had been importing foreign papers for several decades, and many northern European countries would continue to do so, as linen rags were less readily available than in southern Europe.

But it was Italian papermakers who dislodged the Arab competition, undercutting them with the cheapest papers they could make, aided by better mills and more abundant water supplies. (The increased export of wool to the Middle East, especially from England, also meant linen rags became less widely available in the Caliphate's heartlands.) The ingredients were cheaper too, since in the late Middle Ages Europe had begun to cultivate hemp and flax in bulk, excellent raw materials for paper. Europeans used mallets rather than grindstones to pulverize the rags and added iron covers to the mallet heads too, increasing their crushing power. Where the Arabs had used vegetable glues to make the pulp sticky, Europeans were able to improve the consistency by adding animal glue and gelatine. In short, European paper had become both lower in cost and higher in quality than paper manufactured in the Middle East.

Civil servants, clergy, merchants and the literati all profited from the ascendancy of paper in Europe and men of letters were able to become their own scribes, rather than employ others. From the twelfth century, paper was used in parts of Europe for government and commercial documents, and in the thirteenth century it was used for accountancy, private letters and books. By the late fourteenth century paper was decisively subduing parchment and vellum across Europe, and Italy was the chief supplier.

European mills used a technique similar to those in China and the Middle East. As anywhere, they needed a ready water supply. But the earliest European paper mills used old rags as their raw material, so they also profited from being near to linen manufacturing centres like the Vosges region in France. First the papermakers sorted the rags to remove tougher fabrics, then steeped them and left them to ferment. The workers then took the raw rags to the mill, often a converted watermill or corn mill, where small wooden mallets hammered them down to a pulp. Sometimes nails or knives were attached to the mallet heads first. After the beating, the pulp was fed into a vat of warm water and sifted out with a framed mesh; a *coucheur* (paper-layer) would peel off the sheet and spread it across absorbent felt to drain off some of the water. The sheets were then squeezed tight under a heavy press and put in a hanging room to dry. After they had dried, the sheets were coated with size, the glaze used as a filler to give smoothness lest they turn out too absorbent, like blotting paper. After a buff finish, they were gathered into quires of twenty-five sheets. Bundles of twenty quires, five hundred sheets in all, were then ready for market. The process, despite such European additions as the mallet head nails, was essentially unchanged from the one Cai Lun had employed in AD 105. Chinese papermakers had passed their method on to the Arabs. Now the Arabs, thanks to their mills in Spain, had taught the technique to Europe.

These recycled rags were the carriers of a new era of reading in fourteenth-century Europe. Indeed, as paper production blossomed, some countries had to ban rag exports and the growing shortage led papermakers to seek out other products too. (Wood was not used for several centuries; the first wood-pulp book would not appear until 1802.) The Low Countries were using paper for their administration by the end of the thirteenth century and England soon followed. The municipal account books of Mons and Bruges in Belgium were made of paper at this time. Some aristocratic courts had picked up on it as early as the 1270s, but the city-states of northern Italy led Christian Europe in turning out manuscripts on paper regularly, beginning in the 1280s.

Europe's fourteenth-century papers almost always carried Latin on their surfaces rather than vernacular languages. As for the content, it tended to be practical. Thus, there were books on astronomy and medicine, just as there were reference works like dictionaries and legal books. By the last quarter of the fourteenth century, vernacular works were choosing paper too and almost all were devotional. Yet monasteries still preferred parchment for their manuscripts, only using paper for their autograph (or prototype edition) copy, which they would usually throw away later. Nobody seems to have expected paper to last.

By the end of the fourteenth century, however, paper across Europe was five times cheaper than parchment. Moreover, the switch to cursive handwriting had increased the speed at which scribes wrote; they could now deliver two to three folios (four to six pages) a day, a signal improvement on the old rate of just one a day. For the transfer to paper reflected the needs of writers as well as of readers; the switch did not yet turn books pedestrian, but it did allow a rich man's luxury to become, say, a merchant's luxury too. As paper and cursive script brought the price of books down, they began to tap into a market less familiar with Latin, and so vernacular texts became more common.

The mock-knight who sat at his desk overlooking the Thuringian forests in 1521 was a child of this emerging paper age. His learning and languages (he spoke German and Latin, and read Greek and Hebrew), his familiarity with scholars and debates across Europe – contemporary as well as classical – and his ability to study texts for their literal meanings was the only sense in which he could, at the time, be called a Renaissance humanist – a title he would never be given today. In early sixteenth-century Europe a humanist was, in practical terms, an editor of texts. The son of a miner, George had benefited from the economic emergence of northern Italy and its pursuit of learning, which had galvanized so much of Europe, sparking a Renaissance in northern Europe too.

After plague had devastated the Italian countryside, upsetting its medieval order, the rise of Italy's northern cities quickly gathered momentum through the fourteenth century. Cities like Florence and Venice developed into the commercial and intellectual hubs of Europe, trading goods and ideas across and around the Mediterranean. The pursuit of knowledge and beauty became their own ends and this was especially felt in the study of history: once considered a largely spiritual endeavour, history was now increasingly studied for its own sake. Linked to this, the Renaissance allowed for a different perspective on the world; not simply the perspective of God overseeing spiritual history, but increasingly the perspective of the human participant as well. As for the viewer, so for the viewed: the Renaissance also allowed for a greater range of subjects, in both painting and writing, and thus the individual human subject gained an innate dignity and value that is often contrasted with the more exclusively spiritualized focus of art through much of Europe's Middle Ages. Moreover, the Renaissance was not simply some grand shift in meta-narratives played out in statues, in art and on the page, but it was also very much a social phenomenon. The growing interaction of printers and binders and booksellers and readers reflects how the Renaissance was a process – that is, a process of producing texts (their ideas and their formats) – as well as a shift in worldview.[72]

Yet in so far as the Renaissance did represent a break with the worldview of the Middle Ages (and how much it did so is disputed), the Renaissance was, of course, also a self-conscious reconnection with classical Greece and Rome, especially through the media of ancient texts and ruins. This was visible in several fields, none more so than architecture (which we will return to), surely the dominant art of the Renaissance and one which, although the finished product was not on paper, nevertheless relied on paper in its preparatory processes. However, in other fields the importance of paper was more direct, and in all of them it was the Renaissance interest in antiquity, in philosophy, in translation and in the arts which provided a new spur to both reading and writing.

In 1397 the Byzantine scholar Manuel Chrysoloras arrived in Florence (and others would follow) to deliver lectures on classical Greek texts. In the process, he planted the study of Greek firmly in Europe's most fertile intellectual soil. At the same time, Arab translations of Greek works had been making their way to southern Europe, whether into the 400,000-volume library of the Spanish caliph, to the cities of northern Italy or to the continent's leading university, the Sorbonne in Paris. These works were now retranslated into Latin or even into vernacular European languages. Other Byzantine émigré scholars began to arrive in the northern cities of Italy, bringing still more Greek texts with them.

In 1444 Cosimo de Medici established Florence's first public library and began to cultivate a crop of Latin scholars in the city. He believed that the study of Plato could purify Europe's Christianity of its moral depravity. Marsilio Ficino, one of the men de Medici patronized, began to elevate Socrates to a kind of sainthood, arguing that men like Socrates and Pythagoras had been so in tune with moral law that they had been saved through Jesus Christ too.

When the Turks conquered Constantinople in 1453, renaming it Istanbul, still more scholars of ancient Greece fled to Italy's burgeoning northern cities. Indeed, Junker George believed that God had ordained the conquest precisely so that Greek scholars could bring their learning, and especially their understanding of ancient Greek, to southern Europe.

The incursions of books from Byzantium and the Middle East, especially of scientific treatises and classical Greek philosophy, led to a rise in copying in universities across Europe, as scholars sought both to reclaim Europe's classical past and to profit from the new scientific discoveries and ideas to aid it in the future. The spate of borrowings that followed fuelled the growth of reading and writing in the great cities of northern Italy and they increasingly looked to paper as their medium. Paper could stimulate production at affordable rates and in record time. In fact, if there was a bottleneck in book production, it lay with the scribes in their scriptorium, and there was no easy solution to their slowness.

In the 1450s, the Florentine bookseller Vespasiano da Bisticci needed the labours of forty-five scribes and a working window of twenty-two months to copy out a commission he had received from Cosimo de Medici for 200 volumes. This was a rapid rate in a medieval manuscript culture, but Renaissance Italy had a marketplace of readers looking for books about the latest ideas and seeking out the most recent translations from the Greek. It was at least possible to train secretarial staff afresh and, as a result, the status of the writing master grew rapidly through the fifteenth and sixteenth centuries. New handwriting developed as Petrarch,[73] Boccaccio and other early humanists turned back to the Carolingian manuscripts of the classics and imitated not only their prose but the clear style of their script too, replacing Gothic with a quickly written, clear and angled script they called 'Humanistic Cursive'. (Cursive, or joined-up writing, is of course much faster to execute.) In other words, many of the letters were joined-up. In time, Humanistic Cursive script was standardized and the scribes of the Papal chanceries adopted it in the mid-fifteenth century themselves: it achieved a better black–white equilibrium on the page and aided silent reading, in both cases because the letters were less bunched than in Gothic script. It was renamed 'Chancery

Cursive', then 'Italian hand' and, finally, 'Italic', not to be confused with the non-cursive *Italic* form now used by word processors. (Perhaps most notable was Renaissance Italic's replacement of the traditional 'a' with a more efficient form: *a*.) This triumph of speed and clarity in scripts pointed to how the book was evolving too, laying still greater emphasis on message and author over scribe and beauty.

Despite these advances, it was handwriting, not papermaking, which remained the slowest element in the process of making a book. Thus the greatest development for paper came not from the scribes of Italy but from a goldsmith and publisher in Germany. It came in the form of a press. Since the 1420s, Europe had been practising block printing, the technique whereby an entire page of writing, illustrations or both was carved onto a block of wood and, by applying ink to the wood, printed onto a page. But carving a fresh wooden block for each page of a book made production slow – and it could often be better to use a scribe for pages of text.

In the 1040s Bi Sheng, a Chinese official, had invented movable type print, a system whereby each Chinese character was separately carved onto its own clay token, which was then slotted into a metal block alongside other characters. That block was then used to print a page, so you could avoid carving a fresh woodblock for each page of a book. It was an ingenious invention, but Bi's use of clay made it fragile and impractical for rapid-fire, repeated use. (A metal version developed in Korea in the thirteenth century would be more effective.) Bi had also invented it for the wrong script. There are thousands of Chinese characters, which made his invention less of a time-saving device than it would have been for an alphabetic script with its small number of letters. His invention soon became little more than a historical curio.

European scripts, however, were a different matter. Indeed, if movable-type printing could be harnessed to any alphabet-based script, then the results would surely be dramatic, allowing mass-printing at an unprecedented speed. This link was finally made when Johann Gutenberg founded a printing office in the German city of Mainz. Gutenberg was a goldsmith, adept at smelting and forging metals into a variety of shapes. It is possible movable type was his own idea but it may have simply travelled west (as a practice or even as a concept), conceivably with the conquering Mongol armies when they reached the gates of Vienna in the fourteenth century. Either way, Gutenberg undoubtedly introduced it to the European market in the mid-fifteenth century. Yet equally crucial was Gutenberg's development of the process itself, not just introducing movable-type printing, but combining it with oil-based inks and the use of a wooden press.

One difficulty for Gutenberg was that the hard surface of European papers was fabricated to suit tough quill pens rather than gently stamped images and text. He couldn't ask the paper mills to prepare paper specially suited to his new technique, lest he release the secrets of his technology. Thus the only way Gutenberg could hope to make a clear impression on the hard rag paper pages of Europe was with considerable force. His printing machine could be neither a stainer nor a roller nor a stamp. It had to be a press.

The dimensions of European papers determined the size and efficacy of the press too. The fact that European papers were thick, unlike Chinese papers, led Gutenberg to print on both sides, which quickly became standard for European printed books. The major differences between printed books in East Asia, the Middle East and in Europe were the direct result of the type of papers with which their respective book men had to work.

Fundamentally, pressing and printing were nothing new. People had printed images and even words onto fabrics, clothes and sometimes buildings or walls for centuries already, as well as onto wax or parchment with their personal and official stamps. As for presses, European vintners had used them in winemaking for centuries and they were even used in papermaking to squeeze the fresh sheets dry after maceration. Printing on paper was important in European history not because printing or presses had suddenly been invented or discovered but because printing had been adapted, with the press, to paper and, with movable type, to Europe's alphabetical scripts.

Movable-type printing, then, is very much a technology of the paper page. Had Europe been using only vellum, parchment and papyrus, printing could never have enjoyed a fraction of the impact that paper allowed it, given the considerable cost of the other materials. Besides, it is more difficult to manufacture a vellum or parchment surface smooth and absorbent enough to suit quick-fire printing. What followed, therefore, was entirely a paper-made revolution.

Gutenberg began with various different printed works, including several church titles (and probably some schoolbooks), but his greatest printed product was the Bible in Latin. A few copies were on vellum but the paper on which he printed the remainder, shipped to him from the Piedmont region in the Alps, is nothing less than remarkable. Several copies survive entirely intact and his first Bibles can still be leafed through today. It is hard to find such durable paper anywhere. Gutenberg's own contribution was, of course, movable type, and here his crucial innovation was the use of a fount, or set, of type, because it saved countless man-hours of labour. The fount was cast by a punch with a relief pattern at its end made from bronze or brass. This punch was hammered into a small block of softer metal – first lead, in later years copper – to create the matrix, a hollow image of the symbol. To cast a type, you had to clamp the matrix in a steel mould that usually had wooden edges. The type-caster then poured a molten alloy (usually containing lead and tin) into the mould. It solidified very quickly. At that point, you could pull it out of the mould and it was ready for use.

Until you did use it, however, you stored it in a letters case. The case held capital letters at the top and the ordinary letters at the bottom – from this we received the terms 'upper case' and 'lower case'. The compositor then gathered the type for printing, a process called composition. He placed the source text over the *visorium*, or case clip. Then he slotted a line of type into a composing stick, essentially a hand-held tray of adjustable length. When he had filled the composing stick he transferred it to a galley or oblong tray, inside which a reversed image of the planned page would gradually take form.

A sheet of paper was far larger than the book it was supplying, so the bookmaker folded it up until it reached the correct dimensions. If it was folded once it became two sheets, or leaves, and was called a folio; if folded twice to become four sheets it was a quarto; if folded three times to become eight sheets it was called an octavo; and if folded four times to make sixteen sheets it was called a sextodecimo. (The names are taken from the relevant Latin numbers – double-sided printing meant a folio would have four pages, a quarto eight pages and so on.) But the paper was printed before it was folded up and printing would only usually begin when enough galleys had been filled to cover one entire side of a sheet in print.

The press itself might stand six feet high and its central feature was a large screw-thread spindle that closed downwards onto the platen, a bronze or brass flat-bottomed block that pressed onto the page beneath it. The spindle was operated with a wooden-handled iron crossbar. The carriage assembly itself included a shallow wooden box, called a coffin in English, and the galleys were placed inside it.

Two men operated the press. The first man was the inker, who beat the type with a pair of ink pads wrapped in leather and stuffed with wool, horsehair or dog-hair, ensuring the letters were always well-coated with ink. (The chosen ink was probably the same highly concentrated ink later used in Dutch Master paintings.) The second man fitted the sheet to be printed in turn and then swung the central part downwards. Next he turned the spindle handle anti-clockwise to squeeze the galley of inked type down onto the blank page. In this way, two men could produce up to 200 sheets a day or, in the most advanced Italian presses, up to 400. On exceptional occasions in the late fifteenth century a single press managed more than a thousand in a single day, as for a 1481 edition of Dante's *Divine Comedy*. But there were other rapid print runs too in the 1470s and 1480s – Latin Bibles, other religious texts, a commentary on Avicenna and Latin poetry collections.

After the pages had been printed onto, they would be hung up to dry, and then all sheets would be folded in two once they were dry, creating the rudiments of the spine. Of course, such books were then bound but binding was a separate process, as well as an opportunity to beautify the book. Thus the purchaser would receive the printed (or written) sheets and organize the binding separately. This involved sewing each gathering into strong vellum or leather strips, strips that then formed the inner part of the book's spine. The pages could then be bound soft (with vellum or parchment) or else as hardbacks (with pasteboard – a strong, thick paper). Small holes in the outer binding allowed the gathered pages to be sewn into their backing, before a leather or pig's skin cover could be added, and the title usually written not on the spine but the fore-edge (the long, thin side of the book that is opposite the spine).

Setting up a print shop was a high-risk, expensive venture, running into the hundreds of florins. It required printing equipment, paper, labour, premises, distribution and sometimes an editor and translator too. The press itself was a minor cost and could even be hired, but the price of paper could vary wildly, although it usually cost around the same as the labour needed to print onto it. There were many risks (including flood and fire) and you could usually only hope to serve a local market, yet from the middle of the 1460s presses proliferated – first in two further German cities, then in the vicinity of Rome, and later in Venice. In 1470, a press was set up in Trevi and the following year more appeared in many more Italian cities, a trend that continued in 1472. By the end of the century, European presses were set up in nearly eighty cities across Italy, more than in Germany or France, but most of these were commercial failures. The upfront costs required, the need to sell to a much larger market and the general novelty of the whole enterprise (which required a very different distribution model to the old one of manuscript production) meant that the printing trade suffered a high casualty rate. Nevertheless, by 1550 one in eight Italian writers was a printer-publisher. Print brought the voice of the author and the demands of the market closer together – there was no scribe intermediary any longer and print runs usually had to gauge demand in advance. Book production had largely left the scriptorium, where it had been based for almost a millennium. Among the longer-term beneficiaries of this shift would be women and sometimes children, since they might well help their husband or father in his print shop, and so become familiar with the processes of book production and, crucially, more familiar with the printed word itself. A print shop, unlike a scriptorium, could be a family affair.

Print was the offspring of the Renaissance, a technology born of an era intrigued by, even obsessed with, texts. Reading no longer needed to be merely an elite game and books themselves could increasingly be afforded by less privileged men, and even women. Truth itself was increasingly to be sought in studying texts, often classical texts, rather than in appealing to a higher religious institution. Books were read in the vernacular, rather than Latin and Greek, and the subjects ranged widely. Geoffrey Chaucer's *Canterbury Tales* (especially the 1476–8 Caxton press edition) and the Spanish proto-novel *La Celestina* (first published in 1499) both sold well, but so too did histories of the world, among them the Nuremberg Chronicle, published in Latin and then German (both in 1493) with German print runs in the high hundreds, and *Fasciculus Temporum*, written by a German monk, first published in 1474 and running to almost forty editions during the author's lifetime.

Few early humanists gave any indication of actually confronting the Church, but the humanist approach to learning was already a tacit challenge to the Roman Church because it treated reason as the means to knowledge. Few humanists were willing to follow this approach through to its logical conclusion, either from fear or simply because it represented such an almighty change to the world they still inhabited. But the threat was there. Texts were to be studied, written in the vernacular and assessed. Even the Bible itself could be studied not merely in Jerome's Vulgate version but in the original Hebrew and Greek.

The Renaissance, as much as the Reformation, struck at the long-dormant question of authority. In doing so, it began to loosen the strands that still just about held Christendom, or the illusion of it, together. It was slower, more subtle, less daring and less concerned with ideals than the thrust of the Reformation but, like the Reformers, it was claiming new sources of authority. The two were part-and-parcel of the same 'return to the sources' movement that characterized sixteenth-century Europe.

Junker George knew the 'return to the sources' movement first-hand, since he was very much an adherent. In many cases, that meant reading Latin and Greek texts but, in the case of the Bible, of course, it meant discarding the Latin translations in favour of the Hebrew and Greek originals. To make the texts accessible to the new reading public, however, it would also be necessary to translate them into Europe's vernacular languages. And it was while he sat at his desk in Wartburg Castle in 1521, that he began to work on what would become *the* manifesto of opposition to Rome, a translation of the Bible into German. His simple wooden room with its views of the Thuringian hills was the engine-room of the Reformation. In ten weeks he had translated the entire New Testament. The Counter-Reformation would be launched from the marbled halls of Rome, but Junker George launched his revolution from a fortified castle protected on a rock, surrounded by dense German forests and lying at the very centre of Europe, as isolated as it was majestic. It was an apt base from which to proclaim his message to the continent.

By March 1522 George was done with hiding. He left the castle, left behind his knightly robes and even dispensed with his name. He had decided to return home, to return to his old job as a local pastor, to return to the public eye and thus to reclaim his old name of Martin Luther.

Luther was known across the continent in 1522. He was, oddly enough for a pastor-professor in a minor German town, as famous as the Pope, and Europe's first mass-media celebrity. There were many reasons for his fame, from German politics to the writings of the apostle Paul, and from Luther's own prose and beliefs to the spread of print on paper. But it is Luther's own story that runs like a fault line through the tremors that shook and divided Europe in the first half of the sixteenth century, tremors that closed the book on Western Europe's longstanding religious and social order, but which also helped to deliver, at the cost of great blood and pain and division, a Europe better equipped to question its rulers, its institutions, its past and itself.

Luther was born in 1483 to a mining father who had married into the professional classes. But his memories of childhood were largely of a struggle to get by. Once his parents became convinced of the strength of his intellectual ability, they decided to invest in his education. At school, however, he had his first encounter with the weak state of scholarship in much of Europe; he later complained that he was taught very poor Latin. Renaissance influences were yet to reach his hometown and Latin had been weighed down by centuries of scholasticism. Yet it was these influences that would enable him to use paper so powerfully.

Luther was sent to study law at Erfurt University, home to the best law faculty in Europe. Here, he later wrote, he was a regular at the taverns and whorehouses, but he also became known among friends as 'the philosopher'. Here also, as he read the sermons printed cheaply enough for even a student to afford, he became convinced he was a sinner in need of forgiveness. It was during a thunderstorm on a summer's day in 1505, on the road to Erfurt, that Luther, terrified by the lightning strikes all around him, fell to the ground, cried out to St Anne to save him, and swore he would become a monk.

His father was far from impressed. Monks, to Hans Luther, were lazy parasites with a reputation in society at large for whoring, drinking and unearned riches, men who preached poverty but lived in luxury. Yet Luther's excesses as a monk tended towards asceticism and industriousness, part of his personal quest to find forgiveness for his sins and peace with God. Luther's failure to find either of these left him in constant spiritual torment, until his mentor told him that his works could never reconcile him with God; only the death of Jesus Christ was capable of that. He suggested Luther read the Bible and the works of St Augustine.

Luther's breakthrough came as he read the letter of St Paul to the church in Rome, written in the first century AD. Where the medieval Church had developed its own laws and means for sinners to win forgiveness, Luther found Paul arguing that man could do nothing to save himself. Instead, Paul had written that salvation was God's gift, not a good man's wage:

> For in the gospel a righteousness from God is revealed, a righteousness that is by faith from first to last. (Rom. 1:17)

Righteousness was the state of being found just and therefore not guilty of any sin. Paul's argument was that, through Jesus, God had been the author of a topsy-turvy religion: he had switched man and God on a cross outside Jerusalem. A man could only be righteous, Paul argued, because Jesus had made him righteous. Luther himself wrote:

> You will find peace in Him alone and only when you despair of yourself and your own good works . . . he has made your sins his own and has made his righteousness yours.[74]

Justification by faith alone, as this teaching became known, could still just about be plucked from the granary of church doctrines accrued over centuries of European religious history, but it had been both diluted and contradicted by other teachings – and thus largely crushed under their weight. Yet if you did not need a priest or the sacraments or the Pope or even membership of the Church in order to be saved, then Rome was no longer the keeper of the keys of heaven and hell. Images of hell above church doors across Europe would lose the power to undercut anti-clericalism if the Roman Church was no longer the exclusive route to God.

Luther was not the first to oppose Rome over such issues – in recent centuries both the Waldensians in France and Wycliffe's Lollards in England had done just that. Moreover, like the fourteenth-century John Wycliffe, Luther was just one pastor-theologian in one corner of an expansive religious empire. He might disagree with Rome and even openly oppose her on points of doctrine or practice, but to what end? Opposition would mean nothing if the Church quashed the debate or disciplined him. Luther had no army, no political power and he worked in an unimportant German town. He was an unknown with no levers to pull.

Except, of course, for one. And even Luther had no concept of the earthquake – political and social, as well as religious – which deft pulling of that one lever would set in motion.

One of Luther's fellow men-of-letters in this period was Desiderius Erasmus, the Dutch figurehead of the Northern Renaissance. In 1511, Erasmus wrote a scathing satire of Europe's religious life called *In Praise of Folly*. In it, he ridiculed monks' obsession with money and sex, the worship of saints, the weight of scholastic learning and even the overlords of the established Church. Magnanimously, Rome let it pass. In 1516, Erasmus produced a Greek New Testament, drawing on the very earliest documentary sources. In publishing it, he laid bare the mistakes made in the official Latin version of the text. In his Introduction, Erasmus even imagined a world in which the Bible could be read by all men and women for themselves and in their own tongues – '. . . not only by Scots and Irish but also by Turks and Saracens':

> I wish every farmer to sing snatches of scripture at the plough, for weavers to hum phrases of scripture to the beat of their shuttles, and for travellers to lighten their loads with stories from scripture.

Erasmus and Luther had much in common, not least their focus on Greek. Luther's Wittenberg University was only the second university in Europe with a Greek chair. And, like Erasmus, Luther opposed indulgences, the practice whereby any member of the laity could pay off some of their sin by making financial contributions to the Church, thereby ensuring a shorter stay in Purgatory, the posthumous land of limbo which, following a period of evolution, had formally entered Roman teaching in the twelfth century. In 1515, Pope Leo X issued a Papal Bull (or command) declaring a Plenary Indulgence, under which all men were expected to make a contribution and the money would be used to fund a new basilica for the chief cathedral in Rome. Leo's ambition was to build Europe's most dazzling church, but he lacked budgeting skills and he spent much of the money he raised on his own lavish lifestyle.

When a Papal indulgence salesman, a Dominican friar called Johann Tetzel, set up shop in Wittenberg in January 1517, Luther had already been preaching against the sale of indulgences for several months. His chief concern was not with Rome but with middlemen like Tetzel, but he also opposed the concept behind indulgences, namely that salvation could be bought or earned. He sent a list of topics for debate to fellow clergy, a common practice, simply calling them *Ninety-five Theses*. His imagery could often be powerful and pointed, as in Thesis number 50, but he still expressed hope in the pope himself:

> Christians are to be taught that if the Pope knew the exactions of the indulgence preachers, he would rather that the basilica of St Peter were burned to ashes than built up with the skin, flesh and bones of his sheep.[75]

Unsurprisingly, nothing happened. No opponent came forward to debate Luther's points and all Luther had by way of reply for several weeks was silence. But Wittenberg's printing press broke the silence, as unauthorized copies began to circulate and then to travel – Luther also sent copies to friends in other cities, which is probably what sparked printings beyond Wittenberg. It ensured that the Theses were known not just in Wittenberg but throughout Germany and, within weeks, across Europe too. Undoubtedly, such theological discussion points as Luther addressed would ordinarily have interested only theologians, but the *Ninety-five Theses* fell somewhere between high theology and popular manifesto[76] – and they certainly struck a chord beyond the ecclesiastical elite, thanks perhaps to their emphasis on dealing with personal sin and guilt.

Erasmus wrote that in 1520 there were already three Latin editions of the *Ninety-five Theses* and one German edition. In 1521, there were six Latin and another six German editions. By 1523, the Latin version had been reprinted several times and two Dutch editions had appeared. In the year 1525, French and German editions were published, and one Latin edition was published for each of the years 1525, 1526, 1527 and 1528. A Spanish edition appeared in 1527 and two more in 1528. In 1529 three Latin editions and one French edition appeared. A Czech edition soon followed. Although these were not for popular consumption, they did arrest the interest of Europe's elites, and in the process Luther won many humanist scholars and statesmen to his cause.

As early as 1518 the leading English statesman, churchman and humanist, Sir Thomas More, was reading the *Theses*. (Erasmus mentioned Luther's name to More in a letter sent the same year.) Yet even then the Pope was satisfied with merely asking the vicar-general of Luther's monastic order to warn the young lecturer-priest to keep his head down. It was Tetzel, whose fellow Dominicans were known as the 'hounds of God', who promised to have Luther burnt at the stake. Printed copies of Tetzel's counter-arguments were then publicly burnt by Luther's students in Wittenberg. Luther opposed their actions but he did take the trouble to write out the ideas contained in his *Theses* in more detail, while still expressing submission to the Pope. Yet his closing words, which had said there was 'certainly no chance of a recantation', not only angered Tetzel, who determined to have Luther arrested, tried and condemned, but roused the Papacy too. Suddenly Luther found himself at the centre of a storm and, by now, all of Europe was watching.

Following his rise to fame, Luther's first paper outpouring was not printed but took the form of exchanges of letters, as students, scholars and churchmen across Europe began to write to ask him his views on a range of issues. Printing had increased the use of private postal services and paper mail was able to keep a movement's leaders in touch with one another despite a distance of hundreds of miles, and leading reformers and Renaissance humanists all made good use of the service. (In 1505 the Holy Roman Emperor Maximilian I had established a postal system across the empire, and this was the most prominent of the new postal systems. A similar system had already operated among the Italian city-states since the late thirteenth century.)

For Luther, the public airing of his views was a dangerous game to play since it meant his answers came up very publicly against Roman orthodoxy. At a trial in Augsburg, he got away with saying he believed the Pope did not have the power to remit sins being paid for in Purgatory, before being smuggled out of the city by friends. Next he was brought to Leipzig in 1519, to debate Johannes Maier von Eck, chancellor of Ingolstadt University. The judges found a technicality which allowed them to avoid declaring a winner but the debate did air Luther's view that scripture was a higher authority than the Pope and his councils, the doctrine known as *sola scriptura*.

Luther has been remembered by Protestant hagiographers as stout, strong and outspoken, but an eyewitness in Leipzig described him as little more than skin and bone, exhausted by the sudden limelight and pressure. Moreover, some people were already claiming Luther for very different and violent ends, while many Renaissance radicals who had supported him were already beginning to fade away, repelled by the potential personal cost of Luther's ideas. He was now a celebrity throughout Europe and men across the continent began to pin their beliefs and ambitions onto the man who had publicly disagreed with Rome, or with the clergy, or with unchecked power, or with church taxes, or with elitism – there were many agendas people read into Luther's opposition to Rome.

The years 1518 to 1526 were the powerhouse of the European Reformation and Luther's printed writings lay at their heart. A paper-fed Renaissance had spurred the introduction of movable-type printing and the Reformation was now borrowing that same technology. Moreover, printing had much in common with the ideas and ambitions of the reformers themselves.

Printing meant less rote-copying, and therefore less imitation among those who wrote. Its tendency towards exact replication fitted with the Renaissance and Reformation search for literal rather than allegorical meanings. The lower price of its books favoured universal knowledge over elitism, empowering individual readers rather than selected interpreters. Moreover, it identified Bibles as the channels of religion rather than priests and thus encouraged questions rather than acquiescence. It pointed readers back to the author since its content carried very little sign of its middlemen; printers did not seek to personally beautify, interpret or reinterpret texts in the same way that scribes and monks had done. Printing's focus on accuracy supported the return to the earliest Greek manuscripts of the New Testament, and it fostered a read religion rather than an experienced one. The printer himself lived in the city and the marketplace and knew many of his customers, rather than living removed from urban society as a scribe in a monastery.

Printing favoured private reading since for growing numbers of readers its texts were cheap enough to buy and did not therefore need to be read in a library. It was suited to leisure (or simply lay) reading too, since the size of books it produced was smaller than the large-format codices of medieval times, many of which needed a desk to carry their weight. The lower price made the physical object less innately valuable, and reduced the chance of a book being viewed as a status symbol or even as a sacred object in its own right. Thanks to European printing on paper, the author trumped the scribe and, thus, the book's content trumped its beauty and expense. Reading was transformed and the very fact of market-led printing posed a challenge to the unassailability of traditional authority. (It should come as no surprise that it was tricky for the Roman Church to devise a response that was not self-defeating.)

Luther (perhaps unwittingly) harnessed printing's possibilities to a new degree, with a staggering, even profligate output of printed sermons, books and pamphlets. In 1519, Germany printed a third as much literature in the vernacular as in Latin but, by 1521, that ratio had been reversed. Wittenberg, Nuremberg, Augsburg, Strasbourg, Leipzig and Basel became the printing engines of the Reformation, delivering tracts, cartoons, sermons, treatises, theological works and letters for Germany's market of urban readers.

Pamphlets in particular suited the reformers' aims. They were light, easy to carry and conceal, were usually printed in quarto (one of the smaller formats) and only ran to sixteen sheets (or thirty-two pages). Above all, they were not expensive. (One scholar, Mark Edwards, calculated that they cost the same as a hen, a pound of wax or a pitchfork, which suggests they were not dirt cheap but that ordinary men could still afford them.[77])

One study found that 1517–18, the first year of the Reformation, saw a more than five-fold increase in the production of printed pamphlets in Germany, followed by a further eight-fold increase in the years 1518–24. In all, Germany printed more than 6,000 vernacular treatises between 1518 and 1524, running to more than 6½ million copies. In the three decades from 1500 to 1530, there were some 10,000 pamphlet editions (both first editions and reprints) printed in German presses. Most were concerned with the Reformation, and Martin Luther alone was responsible for a fifth of the total, or close to 2,000 first editions and reprints.[78] In addition, between 1518 and 1524 the annual output of pamphlets in Germany jumped by fifty-five times from pre-1518 levels; even after 1524, it still remained at twenty times its pre-1518 rate.

Book printing increased in German lands too, if not at quite the same rate, rising from 416 books in 1517 (110 of them in German) to 1,331 books in 1524 (1,049 of them in German).[79] Mark Edwards found that printings of Luther's works increased from 87 in 1518 to 390 in 1523, before reducing to 200 in the late 1520s. Luther had 1,465 printings (including first and subsequent editions) of his German works, and more than 1,800 of his works were printed by 1525; a further 500 were printed by 1530. In the years 1526–46 he averaged more than three reprints for every first edition. Edwards estimates that from 1518 to 1546, Luther produced five printings to every one printing produced by all the Catholic publicists – even if only Luther's anti-Catholic works are counted, the ratio is still 5:3 in his favour.[80]

But who was actually reading his works? After all, literacy was still uncommon. Moreover, the Reformation was a theological movement, debated among professors and priests and led, in the German states at least, by princes and theologians. Many of them held deep and sincere convictions on, most importantly, the authority of scripture, but how could the Reformation have been a popular movement unless (conversely) it was imposed from above? Even when it did involve popular agitation, it surely remained very far from a movement based on widely read texts. Viewed in this light, the supposed existence of a sixteenth-century 'mass-media celebrity' appears to be a logical impossibility. The historian A. G. Dickens once quipped that historians had believed too credulously in 'justification by print alone' when trying to explain the European Reformation. Has Protestant mythology and hagiography not obfuscated the reality? This is a question that concerned historians of the period for half a century.

Nobody knows the exact literacy rates in sixteenth-century Germany, but urban literacy among men is often assumed to have been around 30 per cent, and at best 40 per cent (although some researchers have argued for higher rates), as against a 5 per cent national average.[81] In short, although literacy was on the rise, and the Renaissance had made reading more fashionable, there had not been a universal reading audience for Luther to galvanize. Although the Reformation enjoyed a very great impact on popular religious practice, on literacy and on politics, it had to be introduced from above.

And yet it is not easy to explain away the scale of printing in sixteenth-century Protestant territories (or in religiously tolerant European lands) if you exclude widespread urban reading from the Reformation. In a conservative estimate, Mark Edwards calculated that Luther's printed output was 3.1 million physical books or pamphlets, yet that does not even include his many editions of partial or complete Bible translations. He counts a further 2.5 million copies of treatises by other German evangelicals and 600,000 by Roman Catholics. In other words, German lands produced 6 million individual pamphlets in a population of 12 million.

Given that Europe's printers were businessmen driven by profit, the fact that evangelicals were producing so much more printed matter suggests not simply that Rome was slow to respond, but also that the market was determining the output. The 40 per cent increase in printed pamphlets in the years 1517 to 1524 led Edwards to question whether sceptical assumptions about literacy rates can really be correct. After all, while Luther's weightier books were often bought up by libraries, princes and clerics (occasionally in bulk), his pamphlets were far more populist and written in a very direct style. Above all, it is the re-publication levels that point to how high demand is likely to have been.

There is no figure available for how many people read Luther's writings for themselves but there are a number of reasons to think that publication levels probably under-represent those who, directly or indirectly, his books and pamphlets reached. There was a large second-hand audience; although silent reading, which amounts to the individualization of reading, was becoming more normal in Europe even before Gutenberg invented his press, it was still common in the ensuing centuries to read out aloud, or to read to a group of colleagues or friends or to your family. (Indeed, this continued across Europe into the twentieth century.) Moreover, Robert Scribner's study of sixteenth-century reading found that German printed popular polemics were in fact designed to be read aloud. Evenings, a common time to read in groups since work was finished, required candlelight, which also favoured choosing a single reader for the group. Any contentious text was best kept out of daylight hours.[82]

Strongly influenced by the church musical culture that had preceded him, Luther wrote hymns by the dozen and in 1524 the first German-language hymnals were printed. His hymns probably had a greater popular impact than his writings on mainstream religion and they were undoubtedly central to religious life for longer than any of his other works, with the inevitable exception of his Bible translation. As with much sixteenth-century reading, hymnals provided a guide for the congregation but it was unlikely they could read every word; often they were simply a memory aid, although familiarity bred a rise in literacy too. Music also presented Luther with a theological opportunity, since he wanted to take the Mass out of the controlling hands of the priesthood and draw the laity, already participants in the experience of the service, closer into its content. Hymns provided a way to introduce a more guttural German into church so as to make the Word of God more memorable and immediate. It was for this last reason that he distributed broadsheets in Wittenberg with words and parts, thereby hoping to enable the laity to learn scripture and its truths through song.

His catechisms had a different but no less profound impact, crystallizing biblical doctrine in a few memorable sentences. Moreover, many Lutherans believed that an education could make them better citizens. The catechism became, for many less literate converts, the sole source of religious knowledge they would read, and even if they could not read it, they could memorize it easily enough: printing could thus communicate to the illiterate too. (The Bible did not replace the catechism as the most widely text in German Lutheranism until the eighteenth century.)

Bureaucrats and schoolmasters would pass Reformation ideas on to their juniors and students too, just as ordinary conversations across the country between the lettered and unlettered would spur the passage of Luther's thought and theology. Before the Sunday papers, most ordinary lay Europeans learnt about politics and news from the Sunday sermon.[83] Thanks to print, clerics now had access to a far greater range of political and cultural updates from across Europe. (Thus print may actually have made sermons more interesting and topical.) The Reformation was very much a preaching revival and it was as they listened to the man in the pulpit that many of the laity probably had their first encounter with Reformation ideas and with Luther's writings, passed on by a reading clergy.

Printing delivered fresh possibilities for images too. The Reformation made use of woodblock printing as well as movable-type printing because its propaganda was formed not simply of letters but of images as well. In the autumn of 1518, Luther met the artist Albrecht Dürer in Nuremberg. Dürer was already an admirer of Luther and a member of a radical Christian group based in the city. He was the artist who, probably above all others, best captured the turbulent spirit of early sixteenth-century Europe. He immediately became Luther's illustrator, and his powerful anti-Papal images and cartoons decorated the reformer's tracts, books and Bible translations – many were graphic, even disgusting. Such images were fundamental to the impact of Protestantism in Europe, often playing with older church imagery and iconography, particularly if it was apocalyptic. But in Dürer's images the familiar scenes were turned upside down, as he depicted Roman Catholic priests being thrown into hell or the Whore of Babylon wearing the Papal tiara, a triple crown.

Dürer was not the only artist who worked with Luther. The portraitist Lucas Cranach the Elder worked for the Elector of Saxony and in 1521 he and Luther published *Passional Christi und Antichristi*, a series of double engravings that compared incidents from the compassionate and holy life of Christ with incidents from the life of a greedy and licentious Pope. Luther himself had said it was a good book for the laity.

Graven images remained important through the early years of the Reformation, but the role of the traditional Gothic cathedral, an encyclopaedia of religious stories in stone and stained glass, was inevitably changed by the rise of lay Bible reading. Moreover, Protestant iconoclasm had strong links to the rise of literacy and, in particular, to the increased accessibility of Bible translations in the vernacular, because they encouraged people away from the sacred icons and rituals of the Roman Church, pointing them instead to the words of scripture on the page. Making the Bible available to all meant that its claims would be comprehensible to all through the medium of words on a page. Ritual, hierarchy and sensory experience were necessarily relegated under the new order. Thus literacy was crucial to the Reformation's success, as paper allowed the message to trump other traditional religious and institutional media.

Moreover, arguments that a reading revolution couldn't have changed sixteenth-century Europe ignore how misleading even the concept of literacy can be. Literacy often means the ability to both read and write, but obviously Luther's print-audience only needed the simpler of these two skills, and not necessarily to an advanced level. One sixteenth-century advert in a German town offered to teach people to read properly, but how were illiterate locals meant to understand it? It is likely that a good number of 'illiterate' townsfolk possessed a very rudimentary popular literacy, recognizing the letters of the alphabet and an arsenal of basic words.

Luther's pamphlets were styled for a popular readership, many of whom were probably only partially literate and who might consult a pamphlet with family and friends, or who simply read their printed matter more slowly than others, sometimes skipping difficult words. Literacy, after all, is not a zero-sum game and printing did hasten standardization, making letters and words far easier for the layman to read. Renaissance printing had already made books and pamphlets more commonplace; the reformers acted as a further spur. As books became part and parcel of the urban landscape, their content (and their script) inevitably grew more familiar too.

Even if only 5 per cent of the German population were reading any of the tracts and arguments being printed across the provinces in the 1510s and 1520s, that was already a momentous expansion of reading beyond the old elite. But it was surely far more than 5 per cent, and far more again who were hearing the content of these works second-hand (and seeing the images which accompanied the texts). Just as sixteenth-century Germany was undergoing a print revolution, it was also staging the continent's greatest ideological dispute for centuries. This shared momentum was no mere coincidence.

It was in 1520 that Luther published two of his most cataclysmic tracts: *To the Christian Nobility of the German Nation concerning the reform of the Christian Estate* and *On the Babylonian Captivity of the Church*. The first of these confidently began:

> The time for silence has passed.

What followed more than matched the drama of Luther's opening line, as he went on to reverse the medieval order of the two estates, placing the secular authorities above the religious authorities (as is accepted across Europe today) and thus playing into the hands of German nationalists and anti-Italian sentiment alike. This was a seismic shift, in line with the apostle Paul's own teaching in his letter to the Romans, but also archly offensive to the established Church. In fact, the work was doubly anti-hierarchical, since in it Luther also argued that the clergy were not above the laity and could therefore not claim moral superiority. All Christians were priests; thus the clergy had neither different political rights nor a monopoly on divine truth. Instead, they must submit to national laws and the people must be allowed to read scripture for themselves.

In the second of these treatises, *On the Babylonian Captivity of the Church*, he was no less forthright. Had it been written just a few decades earlier, it could never have gained much public traction. Luther might have received a warning or some kind of official sanction, but it is unlikely he would have been widely noticed. In hindsight, it is easier to see that Rome's mistake was to underestimate how movable-type printing had altered the public landscape across Europe. Coming down hard on a popular figure whose medium was the printed word would only add to Luther's growing celebrity. Rome could no longer wage its war with Luther in private and on the usual, wholly unequal terms. Pursuing him openly meant turning a troublesome German cleric into Europe's biggest name and meeting him, head-on, through a medium to which he had practically equal access. Rome's reaction to Luther's printed works would allow those works to bring him more fame than Luther could have achieved alone.

No theological debate would be more important than the debate over the Mass. In the *Babylonian Captivity of the Church*, Luther argued that Rome had made slaves of God's people through the ball-and-chain of the sacraments, which had enabled the priesthood to become the means of grace for the laity. Chief among those sacraments was the Mass. The Mass was not solely a priestly performance that took place far from the laity. The vast majority of laymen and women might not understand what was being said or sung in Latin but they could participate in the experience of the Mass itself: in eating the bread, in smelling the incense, in walking through the church as the service progressed, in meeting with neighbours and, above all, in being present during the miraculous transformation of simple bread and wine into the very body and blood of Jesus Christ. The Mass was *the* great miracle of medieval Church life.

Yet explaining that miracle, when the bread and wine remained in their original form, was complicated. Rome had adopted Aristotle's distinction between the 'substance' and 'accidents' of the bread and wine to explain why they did not take on the appearance (or 'accidents') of actual flesh and blood. In the eleventh century the word 'transubstantiation' was used to describe this scientific puzzle and in 1215 Rome formally recognized it at the Fourth Lateran Council.

The early Christian believers had, as in the gospels, gathered round a table to share bread and wine. But, perhaps under the influence of pagan religions, the kitchen and dining tables of the early Church were gradually replaced with the stone altar of sacrifice, as the bread and wine themselves rose in status. Thus official doctrine argued that, when the Mass was dispensed, Christ was sacrificed once again for the believers present. Luther disagreed, arguing that any kind of sacrifice undermined the 'one for all' sacrifice of Christ on the cross. He attacked the use of Aristotle's 'substance' and 'accidents' as 'juggling with words', and he located the power of the event in the faith of the believer, not the actions of the priest. As in other areas, Luther applied his two principles – *sola scriptura* and 'justification by faith alone' – in this case to the doctrine of the Mass. Yet in doing so he had struck at the heart of Church authority. And it was the weight of that authority which he finally had to face when events came to a head in 1521. What followed would spell the demise of Western Christendom as a religious unit.

The 1520 Papal Bull, *Exsurge Domine* (Arise, O Lord), had condemned Luther on forty-one counts of heresy and called for his books to be burnt. Luther's reply had been to publicly burn Papal books himself, throwing the Papal Bull onto the top of the bonfire. He now faced the Diet of Worms (the improbable English translation of the meeting of the German Estates), followed by a likely handover to the Roman Church authorities. In addition to the possibility of swift punishment, the weight of expectation now hung over Luther. Many Germans admired what they considered to be Luther's patriotism, stoked by printed images that likened him to an ancient hero from Germanic mythology. Humanist scholars admired him as the scourge of medieval scholasticism. Moreover, Luther had become an expression and symbol of widespread spiritual frustrations and hopes. As early as 1520, the humanist scholar George Spatalin wrote that nothing at the Frankfurt Book Fair (still the largest book trade fair in the world, it only became a major book fair in the wake of Gutenberg's invention) was bought as often or read as keenly as the works of his friend Martin Luther. As Luther arrived at Worms on 16 April 1521, crowds had lined the streets to cheer him and there were so many spectators that the event had to be moved to the largest room available. Luther must have wondered at the series of turns his life had taken; he was, after all, just thirty-seven years old.

The trial itself was unusual: Luther was asked to recant his own writings and requested a day to reflect first. This was granted and, the following day, he was simply meant to agree or refuse. Instead, he was permitted to speak more fully and, in so doing, he challenged the emperor to take sides between the Pope and Luther. Scripture and 'plain reason' were his authority, he argued, since popes and councils had too often proved themselves unreliable. He refused to go against his own conscience by renouncing his writings. Luther looked set to be delivered to the Church authorities. Instead, Elector Frederick of Saxony had the theologian kidnapped and hidden away in Wartburg Castle, where he went by the title of Knight George.

There was little of the knight about him and Luther's one attempt at hunting was not a success. What he did achieve in his ten-month exile, however, was a translation of the New Testament into German, and it would be hard to exaggerate the importance of this one work on German life. Luther believed that printing was God's great gift to the world, since through its power God would spread true religion to the ends of the earth. In this he was strikingly prophetic – today, the Bible is read in more than 2,000 languages worldwide and the opening salvo of this volley of translations was Luther's printed German Bible. A handful of other German translations were already available, but all of them had been translated from the Latin Vulgate and had preserved many Latin turns of phrase. None were easily readable. Yet by the time of his death, Luther's Bible had been reprinted (in part or in whole) more than 400 times.

He laboured over the translation. Although Luther knew much of the New Testament by heart and had written a cannonade of tracts, treatises and sermons in the German language, he complained that it was fiendishly difficult to reproduce the order of the Greek in his own 'barbaric' language. His answer was to couch the New Testament narrative in prose that reflected his own character: full of energy, texture and brio. Its impact on German literature was as immense as its prose was engaging. This was no less than Luther himself preaching the gospel directly to his readers in language that brought it to life, in German, for the first time. He translated it in just three months.

The Bible was Europe's book of books. For a millennium, while the clergy had debated it, the laity had been denied access to it. Even those who could occasionally access a Bible needed to read it in Latin and were engulfed by a mass of scholastic commentary. Now, finally, many of them could buy the book that everyone wanted to read and they could buy it in prose they could understand and enjoy. Christianity, granted new readerships by the affordability of printed paper, had been allowed to go native.

In his German New Testament, Luther added his own commentary.

Much of this was necessary to explain theological terms that did not translate well into German or simply to take people back to the original Greek or Hebrew sense of words, shorn of their Latin associations. But there was out-and-out propaganda too, with cartoons and marginalia attacking the Pope and his council, most of which equated the Pope with the Antichrist. Indeed, one of Luther's weaknesses was that, having wrested control of religious publishing out of the hands of Rome, he wanted to control everything else. When Anabaptists and Spiritualists began to use this new publishing space for their own ideas, Luther foresaw the inevitable splintering of Protestants across Europe and determined to ensure that the Bible would not simply be left to suffer the exegesis of just anyone. This concern for guidance was perhaps understandable, but either the Bible needed an intermediary or it did not. It was a tension Luther never quite resolved.

Nevertheless, he could at least claim he had placed vernacular Bibles in lay hands and that it was now up to those readers to judge his comments for themselves. It was an undoubted revolution, and the sales figures leave us in no doubt of its groundswell importance. In all its history, paper had never known anything approaching this kind of immediate readership for individual works. The first edition of Luther's German New Testament, known as the *September Testament*, ran to an unprecedented 3,000 copies. It sold out so quickly that in December the printer produced a second edition, with a good deal more editing by Luther. A complete reprint appeared in Basel the same year.

At this point, the New Testaments were in folio format; in other words, they were large, expensive and required a desk. They were aimed at wealthy customers, whether individuals or institutions. However, in 1523 a further eleven or twelve reprints appeared and as many of these were in quarto format as in folio – quarto is half the size of folio and much less expensive. In the translation's peak publication year of 1524, when a further twenty complete reprints were made, most editions were in octavo format, half the size of the quartos – depending on the size of the original sheet, quartos could measure just under 6 by 8 inches. By a conservative estimate, more than 85,000 copies were printed between September 1522 and the end of 1525. The Bible was being tailored to the popular market, its market popularity evident from the decision of printers to invest in reprints. And it was even shrinking to meet new audiences.

Luther's enemies were quick to notice the threat posed by printed papers. One of his Roman Catholic rivals, Dr Johann Cochlaeus, complained that printing was driving Luther's success and that even where his books were banned the inspectors would warn the booksellers of their forthcoming inspection tours in advance. Even tailors, shoemakers, women and the ignorant were, Cochlaeus despaired, learning to read German, sometimes studying under the tuition of priests, monks and doctors. In 1529 Duke George of Saxony (a cousin of Elector Frederick, Luther's protector), lamented that 'many thousands' of copies of New Testament translations were leading people towards insubordination.

Yet the ease of printing paper books fast enough to meet new levels of demand created its own problems. In 1522, eighty-seven different German New Testaments were printed outside Wittenberg without Luther's approval, even though the text, in every case, was merely a bastardization of his own translation. According to one estimate, from 1522 to 1530 four times as many unauthorized versions of Luther's New Testament were printed than there were authorized versions. Sometimes the Papal tiara was removed from the images of the Antichrist (or at least reconfigured beyond recognition) in these editions so as to avoid giving offence to the Papacy. In 1524 Luther even produced his own 'Luther Rose' as a mark of authentication but still the 'copyright' problem continued. The following year he complained that he could not even recognize his own books. (The rise of printing suffered enormously from both its lack of an effective copyright law and from an excess of inaccuracies.)

The sharp rise in output had put quality at risk. As early as 1521, Luther had lamented the amateurish way printing was carried out, from the chaos of the printer's workshop and the lack of care taken by staff to the filthy type founts used and the low quality of the paper itself. Unskilled staff, poor management and the urgency of growing demand all mitigated against accuracy. Moreover, inconstant censorship policies across Europe meant printing businesses were frequently closing down and reopening, which hampered any attempt to develop good working practices. The sixteenth-century printer's workshop was hardly an ideal environment for ensuring the exact reproduction of an original.

Yet through all the mayhem of censorship, intellectual property theft (as we would call it today) and shoddy workmanship, Luther's New Testaments still made it into the hands of untold numbers of readers. The Wittenberg cleric's outpouring of pamphlets and letters also continued. In one letter written in 1531, Luther weighed in on a bizarre debate about the meaning of two verses in the dry Old Testament book of Leviticus. But it was a debate which would lead one of Europe's most powerful kings to break with Rome, adding to the momentum of the paper movement Luther had begun.

14

Translating Europe

I have this peculiar desire to write, even though I cannot decide on subject or recipient. This unceasing desire is for pen, paper and ink, which give me greater pleasure than either rest or sleep. In brief, I ache and decline when I am not writing.

Francesco Petrarch

Luther and King Henry VIII of England had already written on each other's views and it had not been a friendly exchange. A decade earlier, in 1521, Luther had written a critique of the Church's use of its seven sacraments in *On The Babylonian Captivity of the Church*, in which he argued that the priesthood could not be the means of salvation. Henry VIII had replied to Luther's argument with a tract of his own. Helped by his adviser Sir Thomas More, Henry penned *Assertio Septem Sacramentorum*, a defence of all seven of the sacraments. For his loyalty, the Pope rewarded Henry with the title 'Defender of the Faith'. Luther had replied to Henry's defence acerbically:

> Martin Luther, minister at Wittenberg, by the grace of God ...
> Henry, King of England, by the disgrace of God ...

Nevertheless, Henry would later turn to Luther in his quest for a divorce from Catherine of Aragon, his first wife. He was probably less troubled by the doctrine of the Mass than he claimed: the book he took to Mass still carries the flirtatious messages he would write to Anne Boleyn during the ceremony. The Bible, on the other hand, might just offer Henry a means to divorce Catherine and marry Anne. The Book of Leviticus in the Old Testament said a man should not marry his brother's wife, and Catherine had previously been married to Henry's older brother. Henry now hoped to use this in order to have his first marriage annulled.

In 1528 the king gathered almost a hundred books relevant to his case and marked out thirty-seven for them for the royal library. He began to transform his library from a gallery of beautiful manuscripts into a working research library where lawyers and theologians could help him make the case for annulment.

This kind of library was growing more common in England; there were at least 244 significant libraries in the country in the first half of the sixteenth century, almost half of them private. Although the royal palace library at Richmond had only 150 books in 1535, the upper library in Westminster, seat of politics and the Church, had a recorded 1,450 books a decade later. (The universities in both Oxford and Cambridge established small libraries, Oxford in the fourteenth and Cambridge in the fifteenth century, which then began to expand, although in the case of Cambridge significant growth only began in the sixteenth century.) Still, England remained limited in terms of library stocks in the mid-sixteenth century – the royal library at Fontainebleau outside Paris, on the other hand, held 3,000 volumes.

When Henry could not sway Rome, he turned to seek Luther's support for his annulment instead. The cleric turned, as Henry had hoped, to the Bible. But Luther delivered the wrong answer – in fact, he viewed Henry as an irreverent and inconstant self-promoter. Having failed to obtain the warrant he was looking for from either the Pope or Luther, Henry chose, in 1533, to deny the Pope's right to contradict scripture and the following year the Act of Supremacy announced England's break with Rome. And yet it was print, rather than politics, which gave the move real momentum across English society.

Through the 1520s and 1530s, Europe's greatest spur to printing was religion. All the leading English reformers – William Tyndale, Robert Barnes, Thomas Cranmer – were Lutherans, and several were in contact with Luther himself. At the end of the fifteenth century, a number of English scholars had travelled to the continent, especially to Florence, and returned home with much improved Greek, a few recent titles and translations to add to their bookshelves and a new approach to reading texts. Some of the earliest religious humanists, men like John Colet, were in favour of returning to original biblical texts but remained very much Roman Catholic. The reformers of the 1520s and 1530s, however, saw in the Roman Church not so much a series of scholastic study tools and forms of corruption that needed clearing out as fundamental errors in doctrine that denied the claims of scripture. This conviction set them on a much more dangerous collision course with Rome.

William Tyndale was raised in a prosperous family in the Cotswolds, a region which had an evangelical heritage in Wycliffe's Lollards and strong ties to Europe through the wool trade. Tyndale began his Bachelor of Arts at Oxford University in 1515 but was a rarity in a university with few continental students – he knew seven languages as if they were his own. In Oxford he became known as a preacher, perhaps because he was training for the clergy, or perhaps because he used the original Greek New Testament text when preparing his sermons instead of the Latin Vulgate. It was also here that he came face to face with the subjection of scripture to the interpretive lens of Rome. In 1531 he wrote sourly that:

> . . . our holy father proveth the authority of scripture by his decrees, for the scripture is not authentic but as his decrees admit it.

It was in Cambridge, where Tyndale moved in 1517, that he met future reformers like Thomas Cranmer, Miles Coverdale and Hugh Latimer. Here Lutheran ideas were talked about more openly by theologians than in Oxford, yet some viewed Tyndale's learning as a threat. The local Protestant historian Richard Webb recounted that in the Gloucestershire diocese where Tyndale worked as a private tutor after his university studies, in the years 1551–3, 9 clergy were found not to know how many of Moses' commandments there were and 33 did not know where in the Bible they could be found (some suggested Matthew's gospel); 168 of the clergy could not repeat them, just as 10 could not even repeat the Creed and 216 could not prove it, many of them arguing that they trusted Rome and the king for its validity. Moreover, 39 of them could not locate the Lord's Prayer in the Bible, 34 did not even know its author and 10 failed to recite it.

For the laity, the situation was far worse as they could not even understand Latin, let alone read a Bible. Latin had been the only accepted language of Church life in Western Europe for centuries. If parents taught their children the Lord's Prayer in English, there was always the risk that they might be burnt at the stake for choosing their own tongue. Tyndale saw the problem more keenly than most, once telling a priest that his ambition was to ensure that 'a boy that driveth the plough' would know the scriptures better than the priest himself. As things stood, all that was available to ordinary church-goers were gospel harmonies, a spliced selection of gospel stories where gospel content itself was minimal, stories were stretched and fresh narrative colour was added. Wycliffe's fourteenth-century translation of the Bible was in inaccessible Middle English and there had been little interest among senior clergy in bringing the Bible itself to the laity – Tyndale failed to find an official sponsor in London for his translation project or even, as he wrote in the Prologue to his English Bible, 'in all England'.

One reason for this reluctance was the Church hierarchy's fear of Luther's influence. By 1520 there was a flood of Lutheran books and tracts arriving in London from abroad and Tyndale, when he moved to the city in 1523, could hardly have missed them. There were even Lutheran booksellers. The bishop of London, Cuthbert Tunstall, wrote to Sir Thomas More warning of 'Lutheran heresy, foster-daughter of Wycliffe's'. Public bans and government raids followed, but old Lollard networks aided in the distribution of Lutheran works. Weavers, tailors and a whole range of tradesmen were increasingly part of the reading classes, and eager to read Luther's works. Many tradesmen were charged with owning, reading and listening to the vernacular scriptures. A few men were even burnt at the stake for it.

Tyndale left for Germany in 1524 and spent a year in Hamburg before travelling on to Cologne. Here he found a printer, Peter Quentell, and his English New Testament began to come off the press. But he was twenty-two chapters into Matthew when a drunken boast by one of the print-shop workers alerted the authorities to the printing, and a warrant was put out for Tyndale's arrest. He and a companion fled up the Rhine to Worms, carrying their work with them. The beginning of Tyndale's New Testament, printed in quarto format, carried an image of St Matthew dipping his pen in an inkwell; Tyndale must have felt as though he had got little further with his long-awaited printing. But the Prologue, much of it lifted from Luther, began to circulate in England anyway, the first tract of England's Protestant Reformation.

It was in 1526 in Worms, where just five years earlier Martin Luther had made his infamous defence, that a full Tyndale New Testament, in English, was first lifted, page by page, from the press. It was produced in the diminutive octavo format, set in the plainest type and ran to between 3,000 and 6,000 copies. In 1534, when Tyndale was probably still in his twenties, a revised edition appeared, with more than 5,000 changes. This time it was also accompanied by a panoply of woodcut-printed images. The 1534 edition was the first time Tyndale used his own name, although he had to write in his Prologue warning readers of several inexpert (and partisan) edits in the edition anonymously made by his colleague, George Joye. Tyndale later compared what Joye had done to a fox pissing in a badger's hole. There were several pirate printings in Antwerp too.

The beauty of Tyndale's written English lay in its limpidity. He gave new words, word-orders and phrases to the written language that have since become second nature to English-speaking peoples around the world. His choice of words was so striking, clear and stirring that they remain lodged in the Anglophone consciousness five centuries later: scapegoat; am I my brother's keeper; a law unto themselves; man shall not live by bread alone; no man can serve two masters; ask and it shall be given you; as sheep having no shepherd; God forbid; signs of the times; the spirit is willing; fight the good fight. Often he could be direct and colloquial too, as when the serpent tells Eve she should eat the fruit in the Garden of Eden:

Then the serpent said unto the woman: 'Tush, ye shall not die.'

Tyndale's priority was clarity: the teachings of the apostle Paul, the lynchpin of the European Reformation, and the gospels must be understood by the common reader. As a result, he delivered popular English to his readers.

But he had not delivered it yet, of course. He was still in Worms and so too were his New Testaments. Appropriately enough, it was in the pocket-sized octavo format. Thus it carried only the briefest of

Prologues, yet it was long enough for Tyndale to make his Pauline point: the path to salvation is repentance and belief, not works. Justification by faith alone was headed for England and so too was the English Bible, composed in simple, striking prose, printed on paper and bound in an edition small enough to fit in your pocket.

By March 1526 English New Testaments were streaming into England despite the bans quickly placed on them and the public burnings of Luther's books in London. They were smuggled in oil and wine barrels, in sacks of grain, flour or wheat and in the hidden compartments of pieces of furniture. The historian Diarmaid MacCulloch estimates that, during Tyndale's lifetime, 16,000 copies of his translation reached England, then a population of 2½ million with a primitive book market.[84] Some were burnt, others were read and, judging by the well-thumbed copies that survive, probably shared and passed on until many of them began to fall apart. Tyndale had wanted to reach the common ploughboy too and, as early as 1537, Edward Foxe, bishop of Hereford, warned his fellow English bishops:

> Make not yourself the laughing-stock of the world; light is sprung up, and is scattering all the clouds. The lay people know the scriptures better than many of us.

In 1529 the persecutions began under the eager eye of Sir Thomas More, in the form of spying on and investigating suspected Protestants. Publishers and their distributors were particular targets, if their stock was Protestant titles or English Bibles. The persecutions sparked a war of printed words between Tyndale and More. To Tyndale, however, it was the bookburnings, begun in 1527, which were most damaging. Cuthbert Tunstall, the bishop of London, was burning his New Testaments, claiming they contained 3,000 errors. The Church had come to a pretty pass; it was even burning the Word of God.

Wanted by the Church authorities in England, Tyndale travelled back to the continent, where he learnt Hebrew and worked on his translation of the Old Testament, referring all the while to Luther's German translation too. He might have been living in Hamburg, Antwerp, Cologne or Frankfurt. He may even have been in Wittenberg, where he could easily have encountered Luther. Wittenberg, after all, was not only the engine room of Europe's Reformation but of free thinking in general. William Shakespeare's Hamlet attended the University of Wittenberg (founded in 1502), as did his friends Horatio, Rosencrantz and Guildenstern; Christopher Marlowe's Dr Faustus is a resident of the city. Wherever Tyndale was, it was surely in the Lutheran heartlands, a crucible for new thinking in 1520s Europe.

In 1532, Thomas More wrote criticisms of seventeen heretical books; seven were by Tyndale, including his New Testament and his introduction to Paul's letter to the Romans, the most significant biblical book of the Reformation. By 1530 there were reportedly 3,000 copies of Tyndale's *Practice of Prelates* (which contained plenty of criticism of the clergy and the established Church) in circulation in England.[85] A commentary and treatise followed in the next two years. Tyndale was a celebrity in his homeland.

In 1530 England received a delivery of fresh and pocket-sized paper-made arrivals; printed in Roman type, these were copies of Tyndale's translation of the Pentateuch, the first five books of the Bible. Unlike the version produced by Wycliffe, who had translated from the Latin Vulgate, Tyndale's prose was taken directly from the Hebrew. Where Wycliffe had given the medieval 'Be made light, and made is light', Tyndale gave his readers:

> Let there be light, and there was light.

David Daniell, director of Shakespeare Studies at University College London and Tyndale's biographer, has pointed to the impact Tyndale's direct translations from Hebrew had on the English language, from words like 'Passover' to phrases like 'let there be light' to patterns of word order to storytelling techniques. Tyndale himself wrote that he had found Hebrew translated into English 'a thousand times' better than Latin, and that often he had only to translate it word by word for it to read well in his target language.[86]

He continued through the Old Testament to translate the books of Joshua, Judges, Ruth, Kings, Chronicles and Jonah. His feel for English coupled with his understanding of Hebrew made his translations exceptional. Tyndale's Psalms might have been the pinnacle of them all, had he not been betrayed by a hired hand, Henry Phillips, who lured him out of his Antwerp safe house in 1535 and handed him over to the authorities. Nobody knows who Phillips was working for but

Tyndale was found guilty of heresy, tied to a stake in the town of Vilvoorde outside Brussels and strangled to death before being set alight. His dying words were 'Lord, open the king of England's eyes.'

He could never have guessed how quickly that prayer would be answered. In 1537 Henry VIII stated that all doctrine should be argued only from scripture. The lord chancellor, Thomas Cromwell, and the archbishop of Canterbury, Thomas Cranmer, both saw that Henry's comment provided ample justification for authorizing an English Bible across the land for the first time. They moved quickly in their mission. That same year a new English translation had been printed in Antwerp, ascribed to Henry Matthew and called 'Matthew's Bible'. Cranmer wrote to Cromwell, commending the translation as the best yet. Cromwell quickly persuaded Henry to authorize it and in 1539 the 'Great Bible', as it was named, was peeled from the evangelical press of Richard Grafton and Edward Whitchurch in Grey Friars in London. In the 1540 edition (misnamed 'Cranmer's Bible'), the frontispiece showed Henry handing the Bible to Cromwell on his right and Cranmer on his left, symbols of the secular and religious estates respectively. It also carried a new Preface by Cranmer, which not only argued that scripture should be neither wildly interpreted nor withheld from the people but which also recognized the contribution of the Lollards to England's religious life. A copy of the Great Bible was chained to the pulpit of every church in the country and in each parish a reader was ordered to read it out to illiterate parishioners.

The impact of this decision was seismic and irreversible. In 1530 English religion was delivered to the laity by the priesthood. The Mass and its power was passed to those in the pews by priests (and then only in the form of the bread). So too were the other sacraments of the Church. With the exception of some vernacular and 'macaronic' (mixed Latin and vernacular) sermons in the late Middle Ages, the service was performed and delivered in Latin: from the Mass to readings from the Bible. Laymen generally could not speak or understand Latin (save perhaps for a very few common religious terms) and the priests therefore offered the only route to God and to salvation. A priest would interpret the scriptures and his congregation had little recourse to check what he said. Access to Bibles people could read would knock down the largest barrier between the laity and the book they were meant to follow.

The very existence of printed English Bibles in churches already undermined Papal claims to supreme authority, and a conservative estimate of the number of Bibles and New Testaments printed for the English market between 1520 and 1649 puts the total at 1.34 million, more than there were households across the country.[87] But they also began to erode the unchecked supremacy of the priestly elite and to involve the laity in religious beliefs and life far more directly. That change had enormous consequences that were, inevitably, felt in politics too. For if scripture was above the clergy, then it was also above the king.

That was not all. In 1611 the Authorized Version (or 'King James Version') of the Bible became England's officially sanctioned scriptures; it maintained this supremacy for three centuries. Nine-tenths of its English New Testament, and the first half of its English Old Testament, were simply transplanted from Matthew's Bible. As a result, the words, phrases, constructions and style of that earlier translation became a unifying force for the English, a shared word-quiver of spoken and written unity that was so deeply grafted into English consciousness as to become indivisible from English identity. It has been the most influential book in the English language and its translator, who probably used the unknown name Henry Matthew because his own was too provocative, was William Tyndale.

The enormous importance of the Authorized Version is well known, and the reach of its influence is well documented, on English parliamentarianism, the slave trade, and on both the blossoming of seventeenth-century English literature and the ideas of liberty contained in some of those texts, such as Milton's *Paradise Lost* and his *Areopagitica*, which argued for the freedom to print views openly. The principles of exegesis and independent inquiry that popular and personal Bible reading encouraged were also linked to the growth of English empiricism, with all that would mean for the flourishing of scientific inquiry – the study of 'God's other book'.

As had been the case in China many centuries before, the emergence of papermaking in Europe did not automatically translate into popular reading. Even the invention of movable-type printing, a development very much part of paper's own story, could not simply create a wider reading public. Instead, paper and printing provided the opportunity for reading to spread rapidly, should the need or desire arrive. It is for this reason that papermaking's initial arrival in Europe (whether in Islamic Spain in the eleventh century or Catholic Italy in the thirteenth) is very far from being the highlight, or even the turning point, in paper's European journey. It was, rather, the Renaissance and Reformation that provided the impetus for scholars and theologians to go back to original texts and the decision to deliver those texts to the reading public in their own tongue, which acted as an opening of the floodgates for paper to swamp the continent.

For three centuries prior to the Reformation there had already been a number of Bible translations into vernacular languages produced. The fifteenth century in particular had seen not only the Roman Catholic use of printing for everyday purposes – service books, devotional literature and sermon manuals – but also a spate of Bible translations printed across Europe: in German (1466), Italian (1471), Dutch (1477), Czech (1478), Catalan (1492), abridged versions in French (1474) and in Spanish and Portuguese (both before 1500). Moreover, in 1517 Spanish Catholics had completed an enormous project to publish the Bible in its original languages; it is known as the *Complutensian Polyglot* (produced at Alcalá de Henares, 'Complutum' in Latin).

Numerous as these translations were, however, they were unable to galvanize the continent as their sixteenth-century successors did. First of all, they were in largely inaccessible language, translated from Jerome's amateur Latin, the Vulgate version. A linked problem was that they were produced only for the eyes of the clergy, royal courtiers and university professors and students; they were certainly not part of a project to equip the laity with scriptures. Their obfuscatory style only confirmed this; in short, they lacked clarity. The sixteenth-century translations, moreover, had the advantage of fresh biblical scholarship.

Erasmus's Greek New Testament of 1516 was the foundation text of the Reformation since it enabled evangelical scholars to return to the original language of the text. Luther did draw on the writings of other evangelical breakaways – indeed, it was when reading the works of the early Czech Reformer Jan Hus that Luther exclaimed 'we've been heretics all along'. But it was the appearance of Erasmus's Greek text that allowed Luther and other translators like him to achieve their goal: Bibles for the ordinary man, Bibles free of the awkward and inaccessible Latinisms of the past.

These new translations appeared quickly. The Baltic rapidly turned Lutheran: a Danish version of Luther's New Testament was published in 1524 and a Swedish Bible translated from the original languages was published in 1540–41. (By that time both countries had adopted Lutheranism as their national religion.) Bible translations appeared across Europe in the 1520s (Dutch), 1530s (French, German, Italian and English) and 1540s (Finnish, Icelandic). A second wave arrived between 1560 and 1590 (in Polish, Czech, Welsh, Lithuanian, Slovene and Hungarian), with an Irish version appearing in 1602. Europe's book of books was courting local audiences.

Moreover, as Geneva became the nerve-centre of the Reformation a few years after Luther's death (under the leadership of the French Reformer John Calvin) evangelical works began to pour into France too, and fresh Greek and English New Testaments were produced and printed in Geneva itself, which had a significant Protestant English expatriate community in the 1550s. (*The Oxford Companion to Shakespeare* calls the 'Geneva Bible', as it was known, Shakespeare's 'primary text' of the Bible, based on a reading of biblical references in his plays.)

Calvin himself could produce an octavo work of a hundred pages (or 17,000 words) in just a few days, although as a preacher he continually used the spoken word to spread his message, preaching more than 2,000 sermons in Geneva alone, and he often found printers to be unreliable. (There were other risks in the publishing process too: his commentary on the apostle Paul's second letter to the Corinthians was lost on its way to be delivered to the Strasbourg printer Wendelin Rihel in 1546 but he had made no copies, largely because of the cost in time and money it would have required.) We know of 1,247 letters which Calvin wrote and received. Like other reformers, Calvin saw letters as meant for private purposes. Luther's friend and ally Philipp Melanchthon wrote and received more than 9,000 letters in his lifetime and Luther's own tally was more than 3,600. But if Calvin's paper correspondence paled beside his fellow leading reformers, his written output in general was almost fabulous.

The *Institutes of the Christian Religion*, his greatest work, ran to 450,000 words in its definitive 1560 edition and his commentary on Paul's letter to the Romans was 107,000 words. In his exegetical works alone, he averaged 65,000 words per year. Add up each physical copy of each of his works that was printed and the total is approaching four million.[88] Yet that number does not even include his sermons, lectures and correspondence.

Calvin's impact through paper was enormous, not least in France, which he had left in 1536. He reshaped the French language, often proving himself original in his writings where his peers were prolix. Moreover, his ideas became such a threat that in 1551 the French authorities banned French citizens from contact with the city of Geneva and declared that possession of any books published in Geneva would henceforth be taken as conclusive evidence of heretical beliefs. (Calvin's Geneva, of course, was equally ruthless in its attitude to opposing doctrines.)

Europe had printed and preached its way into a continent-wide division, encouraged by political fault lines that ran very deep. By 1545 almost a third of Europe followed an evangelical form of Christianity. At one point even France looked as though it would turn to a more evangelical Christianity too. Lutheranism had offered Bibles and religious tracts to the laity, and Rome's long-standing supremacy was no longer merely precarious – it had been cut back and openly questioned in the strongest terms.

Moreover, later in life Luther wrote more directly on the Papal institution itself, turning even to the documents on which Papal supremacy was based. In 1440, the Renaissance scholar Lorenzo Valla had demonstrated that Rome's claim to historical superiority over Europe's Church had been founded on a deceit. The claim had been made in the 830s by Pope Gregory IV when he issued the *Donation of Constantine*. According to the document produced by Gregory, Rome's supremacy had been asserted by Emperor Constantine himself. Constantine, after a vision before a crucial battle in 312 from which he emerged the victor, had converted to Christianity and lifted the ban on the religion that had been in effect throughout the Roman empire. Later, in 330, he moved the capital of the Roman empire east, to the newly founded city of Constantinople, and then, the document indicated, he had bestowed spiritual authority over Italy and Western Europe on the bishop of Rome.

Valla, though certainly not the first to question its validity, demonstrated that the document was a ninth-century forgery; it was part of the *Decretals of Isidore*, understood to be letters and decrees of the early popes which added weight to Rome's claim to leadership, but these were forgeries too, and also began to be more widely questioned at this time. In short, Rome's badges of historical support for its supreme position in the Western Church were being more openly questioned; the gradual centralization of institutional and political Church power, which had received its first great boost under Gregory the Great (bishop of Rome from 590 to 604) now had its public detractor. Rome's claims had peaked in 1302 when Pope Boniface VIII issued a Papal Bull entitled *Extra Ecclesiam nulla salus*, 'Outside the Church there is no salvation'. Boniface, of course, meant not just any church, but the Church of Rome.

In 1537 Luther attacked the *Donation of Constantine* too, not because it needed attacking – it had been discredited for decades – but because he wanted the Papacy to take the next logical step and decide on its appropriate position within the Church. The ambition of Luther's challenge was extraordinary but it was also proof of how far and how quickly Europe had changed.

Lutheranism had forged a print space for dissension and opposition which was unprecedented. There had been criticisms of Rome before, but they had almost always remained within the establishment rather than finding a pulpit outside it. When Rome excommunicated Luther and failed to reclaim each of the German states under its spiritual headship, a new intellectual space appeared and Luther became its pioneer polemicist, rebel and propagandist.

Luther himself had used this space to attack one after another of Rome's additions to early Christianity: Papal supremacy, transubstantiation, monasticism, indulgences, Purgatory. Fundamentally, however, he had used the space to protest against the belief that a man's own works or his participation in the Church could win him forgiveness and salvation. According to Luther's logic, the laity no longer needed an institution to mediate their salvation; furthermore, they no longer needed an institution to mediate truth.

It was not long until the liberty of expression Luther had achieved found other avenues.

15

A New Dialogue

SCENE II. – *The Forest.*
Enter ORLANDO, with a paper.
ORLANDO
Hang there, my verse, in witness of my love:
And thou, thrice-crownèd queen of night, survey
With thy chaste eye, from thy pale sphere above,
Thy huntress' name, that my full life doth sway.
O Rosalind! these trees shall be my books,
And in their barks my thoughts I'll character;
That every eye, which in this forest looks,
Shall see thy virtue witness'd everywhere.
Run, run, Orlando; carve on every tree
The fair, the chaste, and unexpressive she.

From Act III, As You Like It

Georg Joachim Rheticus was a scholar and mathematician whose alma mater was the University of Wittenberg; here also he would be given a Professorship by Luther's great ally, Philipp Melanchthon. Rheticus friendships with printers meant he had probably learnt how good Protestantism could be for business. But he was also a Renaissance man (he taught mathematics) with an eye for new works that might win him money and fame and, in the 1540s, Rheticus spotted an opportunity. Wittenberg's books and pamphlets were already in huge demand across Europe but he had also seen Luther's own readership shrink from European-wide to just German; the Reformation had begun as a German movement famous across Europe, but, once it had spread, it had begun to lay down local roots. Luther's unprecedented popularity with readers across Europe, therefore, had a limited time-span.

In 1539 Rheticus had arrived in Frauenburg in Poland, travelling the 400 miles north-east from Nuremberg, to study under the great mathematician and astronomer Nicolaus Copernicus. Melanchthon had made the arrangement although he, like many leading Protestants, would write scathingly of Copernicus's great theory. The theory, which defied established wisdom, argued that the earth travelled around a stationary sun (which lay at the centre of the universe), a model known as heliocentrism. This contradicted geocentrism, the accepted understanding that the sun travelled around the earth, with the earth at the universe's centre. Copernicus's idea was opposed by the established Church, but his views had been circulating in manuscript form since 1514. Now Rheticus, after an initial testing of the waters in which he wrote and published his own short introduction to Copernicus's theory in 1540, wanted the theory printed in full. He persuaded Copernicus to publish the work, first in a tentative and reduced form, then in full. Rheticus found a printer for the work in Nuremberg and a proof-reader called Andreas Osiander, the Lutheran pastor of Sant Lorenz Kirche. However, Osiander added his own, unattributed Preface, which argued that Copernicus's intentions had not been to assert 'literal truth' but simply to demonstrate his methods. He also changed the title, *De Revolutionibus Mundi* (On the Revolutions of the Earth), to simply *De Revolutionibus*, thus muddying Copernicus's heliocentrism. The book was printed, in 1543, under a title that offered little more clarity: *De Revolutionibus Orbium Coelestium* (On the Revolutions of the Heavenly Spheres).

The Roman Catholic Church was not against science in principle or, very often, in practice, although European science had for centuries been backward when set against the advances of China and the Abbasid Caliphate. Rome had encouraged astronomy for decades and generally subscribed to the arguments of Aristotelian science. (The Church did, however, reject the Aristotelian teaching, then widely accepted, that the universe was eternal.) Yet it was this same loyalty to Aristotle and, in this particular area, to Ptolemy, the second-century AD Greco-Roman astronomer who had written in favour of geocentrism, which would prove so damaging in the Church's encounter with Copernicus. Moreover, Rome had traditionally interpreted various poetic biblical verses about the sun as supportive of its geocentric view. On this, Luther agreed. Indeed, the Protestant mainstream did not welcome heliocentrism either.

The printing of Copernicus's work was largely uneventful – in Iberia the Roman Church even officially sanctioned it. *De Revolutionibus Orbium Coelestium* had an initial print run of 400–500 copies, an impressive number for a scientific work. The second printing, in Nuremberg, ran to the same number. There was a third printing in Basel. The work even found its way to the major astronomy professors and libraries as well as to several kings, a count and an elector. The greater problem came with Galileo Galilei, the Italian scientist who was condemned by a Roman Inquisition in 1633 for making the case for Copernican astronomy. For Galileo it was unfortunate that his ideas were being broadcast in the aftermath of the Lutheran Reformation and in the midst of the Thirty Years War. Rome, defensive in a time of war and loss, reacted badly. Protestant reaction to heliocentrism was not initially warm either, but it would soon become more varied. Thus John Milton, whose Protestantism extended not merely to opposition to the Church of Rome but even as far as opposition to the very existence of an institutional church and of a church liturgy, visited Galileo in prison before writing the *Areopagitica*, his defence of press freedom; moreover, Galileo's work was printed in the Protestant Netherlands, where press restrictions were significantly looser than elsewhere in Europe. Protestantism's replacement of the supreme authority of the institutional Church with the authority of the scriptures now meant that, in the babble of Protestant voices arguing their different points from those scriptures, there was a good deal more debate and disagreement in general, and this fostered far more space for independent thinking.

When England's Royal Society (the Royal Society of London for the Improvement of Natural Knowledge, to give it its full title) was founded in 1660, one of its opening speakers argued that, just as the Protestants had purified scripture (as well as returning to the original Greek and Hebrew texts, Luther had removed the 'apocryphal' books from the Roman Catholic canon[89]), so Copernicus and Galileo had purified science. Frequent references were made to the two great books of learning: the book of scripture and the book of nature – in both cases, it was the empirical method that made study so fruitful. Isaac Newton's personal library showcases this relationship especially well; among more than 2,000 titles, books on theology and science dominated.

The Copernican print runs could not compete with those for Lutheran tracts or other popular Renaissance writers, but nor were they supposed to. Where the Reformation drew wide audiences and interest, other areas galvanized by print had different aims. The emergence of a scholarly scientific community across Europe, able to read one another's works with ease, was paper-printing's greatest gift to scientific debate and discovery. Printing enabled a communications revolution, after all, and just as Europe's religious unity under Rome was dissolving, so its thinkers and inventors were discovering a new European intellectual community to which they might belong. Scientists still look to a handful of publications on paper as the most important dates in European science up until the early twentieth century, since these publications were quickly being read by scholars and scientists across Europe. The historian Adrian Johns lists the first two of these as Copernicus's *De Revolutionibus* (1543) and Galileo's *Dialogue on the Two Principal Systems of the World* (1632). He includes only a further four publications – by Newton, Lavoisier, Darwin and Einstein.[90] The major scientific breakthroughs were all announced on paper.

This ability to capture more specialist disciplines, such as science, was part of the continued rise of paper culture in Europe during the Renaissance. But niche subjects could never have delivered the paper age on their own; instead, the crucial groundwork was done by the networks of printers that emerged across Europe during the fifteenth century; by 1500 there were printing presses operating in more than 200 cities across Europe in a dozen different countries, producing more than 20 million volumes. Most of these printing presses would be economic failures, leading the print trade to gradually coalesce around a few key urban centres, but the initial surge at least indicated the unparalleled ambition of booksellers in the age of print or, to put it another way, their conviction that the book market was expanding at breakneck speed.[91]

The appearance of these presses also reflected the Renaissance love for the book – and early Renaissance books were often objects of exceptional beauty. Thus Renaissance authors and publishers harnessed the visual potential of paper and books, even after the introduction of movable-type printing. Form was suggestive of content; elegance, harmony, learning, advancement and a love for the classics – even for classical form – could all be signalled through the look and feel of the book itself. Venice became the powerhouse of fifteenth-century publishing, producing a quarter more editions than Paris, its main competitor – and, unusually for the time, more than half of Venetian books were in the larger folio format. Venetian books used the best paper, employed the most impressive designs, blazed a trail in using illustrations and indexes, and yet were also less expensive than their non-Venetian rivals. It was a city comfortable with the economics of books, as print runs required significant initial outlays (and therefore risk). It was also well-placed to export its books across Europe.

Meanwhile, the northern Renaissance spurred paper production in France. One of the first centres of northern papermaking had been Avignon, where the Papacy had relocated from 1309 to 1378. (Following a deadlocked conclave divided between Italian and French cardinals, the French Clement V was finally elected Pope by a narrow margin in 1305 and chose – perhaps wisely, given how completely he owed that electoral victory to the French cardinals – to hold papal court in France, moving to Avignon in 1309.) The humanist popes of Avignon were less interested in piety than scholarship, introducing to France both Petrarch and Boccaccio, each man a forerunner of the Renaissance. By the late fifteenth century, however, Paris and Lyons had become printing centres not just for France but for Europe, producing more printed materials than any other cities in Europe, except for Venice.

If the Renaissance spurred book production and made learning and ideas more available, it was the reformers who first saw printing as a way to reach every reader in Europe – even, perhaps, the illiterate, whether they encountered those texts directly, through images, or indirectly, by hearing them read and quoted aloud. Whereas the Renaissance focus on classical learning (often expensively bound) could deliver it to only a relatively elite audience, religion in Europe was universal. Allied with printing, religion had the potential to reach a mass audience. Moreover, once the Reformation had taken hold, reducing the price of printing, forging new distribution networks and encouraging reading much further down the social scale, it then became more feasible economically for other disciplines to profit from printing, too.

As we shall see, paper was able to capture disciplines that had formerly been closed to it, and to give them a new network of writer-readers across much of Europe. Scholars in such niches were not usually aiming at the broad readership that Luther had sometimes written for, but their more targeted audiences were important for very different reasons. New reading networks and markets would now emerge for composers and musicians, playwrights and artists, scientists and engineers, prose-writers and poets. The result was an emerging culture of cross-pollination, one that was able to deliver moments of radical breakthrough with far greater frequency.

In several of the arts, the links the early Renaissance enjoyed with paper often appear very indirect, at least at first sight. In early Renaissance art, for example, the dominant form was mobile pictures, and among these it was not paintings but altarpieces which dominated. Even when destined for a private home, these products usually comprised wooden panels. Beyond altarpieces, moreover, it was not works on paper which dominated but frescoes and paintings on panels. Innovations in Renaissance painting included the use of oil paints, the employment of 'pagan' subjects and a new understanding of light – compared to these developments, paper in Renaissance painting seems more footnote than feature. Its importance lay, instead, in the role of preparatory sketches and drawings, which were usually on paper or parchment.

Cennino Cennini, writing no later than the early fifteenth century (and possibly earlier), described drawing as the foundation for art and advised painters to make drawings on paper, parchment or wooden panels on a daily basis. His *Book of Art* or *Craftsman's Handbook* was a distillation of artistic techniques and advice. Such preparatory sketches must have been especially important when it came to figurative compositions. Moreover, the idea of preparatory drawings was not entirely new; Petrarch had described them as the common source of sculpture and painting as early as the 1340s. It is difficult to gauge whether paper was used significantly more than parchment or panels for such sketches, since artists often did not consider them worth keeping, although the widespread availability of paper in Italy in the fourteenth century, coupled with its lower cost, suggest it was at least the natural choice for the financially prudent. Moreover, several Renaissance figures valued drawings on paper very highly – not simply the artists we most associate with them today (notably Leonardo da Vinci) but even the sixteenth-century father of art historians, Giorgio Vasari, who was one of the first collectors of such drawings.

Architects had perhaps even greater use for paper than painters, at least in the early Renaissance. They could increasingly afford to buy architectural manuals with recent drawings. Some drew their designs out on what we now call cardboard (*cartone* in Italian); the origin of the word 'cartoon' comes from this use of heavy paper as a surface to make preliminary drawings for frescoes, paintings, tapestries and stained glass windows. These technical drawings, the work not simply of artists but of artist-engineers, would be crucial for the development not only of architecture but of the Renaissance vision more broadly. This was nowhere more apparent than in the work of Filippo Brunelleschi, a Florentine sculptor who quickly switched his trade to architecture. Vasari, in his *Lives of the Artists*, writes that Brunelleschi never tired of making drawings of buildings he saw, sketching them out in their form as in their details; thus he studied and progressed and experimented through imitation on the page. It was as an architect that Brunelleschi pioneered the use of linear perspective in drawings and paintings, whereby the two-dimensional page can be convincingly used to present three-dimensional objects. (The concept behind the practice, aptly enough, seems to have been the brainchild of an eleventh-century mathematician in Baghdad.[92]) Others had been trying this for centuries, but Brunelleschi experimented with mirrors to develop a more precise understanding of how lines converge on the horizon as they disappear into the distance. The result was a more technical basis for representing depth on a two-dimensional surface. Brunelleschi's breakthrough was rapidly welcomed across Florence – indeed, together with his ability to capture technical processes on paper it would be used across architecture and engineering in the centuries to come.

But if paper fuelled the experimentation of architects, it was printing which delivered to them both easier access to foreign designs and a wider European market for their own. The 1550 publication of Vasari's *Lives of the Artists*, which offered brief biographies of several major Renaissance architects, undoubtedly helped to raise their status. As that shift occurred, printing delivered the possibility of making their presence felt in distant countries, since architects and men of letters abroad might now afford to buy their works. Their target was not Luther's mass market but a far narrower readership of potential patrons and men of influence. No Renaissance architect enjoyed a greater impact through the media of paper and print than Andrea Palladio.

Born 'Andrea, son of Pietro the miller', he was later named 'Palladio' by the humanist poet Gian Giorgio Trissino (who employed him in 1538–9) as a reflection of his classicism, since he looked back to the Roman architect Vitruvius and to classical form as his starting point. Almost 300 sheets of Palladio's drawings survive, covering the span of his long working life. Palladio used various sizes of paper, but they were all sourced from sheets that measured around 22 inches by 16½ inches, which he might then divide into two or four sections. He would incise construction lines on the paper with a stylus and then go over them with an ink pen, typically made from a feather. Wooden rulers and brass compasses were features of his everyday stationery, ensuring straightness and correct orientation. According to Vincenzo Scamozzi, his principal successor, Palladio sketched initial drawings with black chalk or a lead pencil. If he had drawn the design for an ancient building on the recto (front) side of a piece of paper, he might then sketch a derivative idea on the verso (back).

Yet paper did not merely aid the planning of his buildings. It propelled his reputation and influence across Europe and even across the Atlantic. Palladio's signature work, *I Quattro Libri dell'Architettura* (The Four Books of Architecture), was published in Venice in 1570. It was unorthodox because, in Book Two, Palladio actually treated his own works. For all its freshness, a treatise on architecture, complete with woodcuts of Palladio's drawings, was not destined to appeal to a large readership, and nor did it need to. The *Four Books* was, typically for Renaissance artists who relied largely on wealthy patrons, aimed at a rarefied readership. There is some evidence that the Italian edition travelled around Europe – the English architect Inigo Jones owned a 1601 edition before his own trip to the Veneto in 1614. But drawings alone did not make Palladio famous abroad. Instead, it was translations which would, decades later, transform the architect's influence.

In 1645 an abridged Paris edition of the *Four Books* appeared, followed by a full-text version in 1650, the first translated from the Italian. In 1663 an English edition of Book One was published, and a complete English edition finally appeared between 1715 and 1720. Various editions followed, none of them accurate and many of them with altered illustrations, until a definitive 1737 edition became authoritative for English-language readers. In 1734 *Palladio Londinensis* appeared on the English market and quickly became the standard builder's manual. A few copies of the *Four Books* travelled to America too, especially to well-connected or wealthy readers like Thomas Jefferson, who in 1816 wrote in a letter to a friend: 'Palladio is my Bible'.

Yet grandly bound portfolios of the *Four Books* aimed at wealthy readers did not create such a stir in America as did the pattern books of Palladio's designs. Pattern books showed models and designs to the reader in a more affordable format. These books had developed thanks not to letter-type printing but to block printing; where movable type had transformed letter-type printing, block printing had profited from improved craftsmanship, specifically in carving patterns in wood.

Pattern books of Palladio's designs and drawings, printed in London, popularized Palladio among gentlemen, builders and craftsmen in America. Meanwhile, the 1734 *Palladio Londinensis* became the most available book on architecture in Britain's colonies. (In fact, its Palladian heritage was impure but by this point many publishers were more interested in the celebrated Venetian's selling power than in achieving the most accurate portrayal of his ideas and drawings.)

Inigo Jones and Richard Boyle, Lord Burlington (the wealthy English lord known as the 'Apollo of the arts'), brought Palladio's drawings to England, thus nurturing new disciples of the Venetian, among them Christopher Wren and Nicholas Hawksmoor.

It was in America, however, that Palladio's influence was perhaps most enduring and monumental. Thomas Jefferson drew up his designs for Monticello, the classical villa at his Virginia plantation, based on his close reading of Palladio's works. (The name 'Monticello', 'small hill' in Italian, may even have been a nod of recognition to the Italian.) Numerous plantation houses and several university buildings in the United States were built on Palladian designs in the eighteenth and nineteenth century; his influence even extended, through the conduit of James Hoban, the Irish Palladian architect whose submission won the contest to design a new building of state in Washington DC, to the design of the White House.

Palladio's use of paper as a starting point for his art was typical of Renaissance architects. It allowed them to test their ideas freely, to engineer new ways of transferring weight or capturing light without ever lifting a stone. Thanks in great part to Brunelleschi's development of perspective drawing, paper delivered architects a means of practising their art far more often than real building projects would have allowed, and with complete freedom. Moreover, the growing availability of architectural texts, coupled with the possibility of becoming well known in print hundreds of miles away from the buildings you designed, delivered entirely new possibilities to Renaissance architects. The buildings are of course the legacy, but paper had helped to make that legacy possible.

Renaissance architects were fortunate to have surviving classical buildings and architectural treatises from which to learn. As they sought to revive the best of classical antiquity, they discovered that much could, indeed, be recovered. Yet that precious access to classical precedents was not shared by composers and musicians. While many were eager to revive what they knew of the classical passion for music, it was not possible to revive the music itself, since the few traces remaining gave very little away of what the music actually sounded like or how it should be played.

In medieval Europe, musicians lacked detailed musical notation until the development of stave lines from the eleventh century – the familiar five-line stave then appeared in Italy in the thirteenth century. Aside from wandering minstrels, music in medieval Europe had been largely based in cathedrals and churches for centuries. But the spread of printed music, begun in the second half of the fifteenth century though only significant in quantity and impact from the start of the sixteenth, linked composers and choirmasters across Europe; thus they could increasingly read and play one another's work without travelling or paying large sums for manuscripts. Yet printing out music posed practical challenges that printing out text did not.

The major problem was how to print notes onto a stave. At first, printers only printed the staves; all the notes were then added by hand. The Mainz Psalter, produced in 1457, is one example of this approach. This, of course, saved some time but left most of the hard work to the scribe. Yet superimposing notes onto the stave required exceptional accuracy and it was complicated by the fact that, in liturgical music, the standard form was black notes on a red stave. To overcome this challenge, printers would impress the staves first, and later the notes, a process begun in Rome in the 1470s. Ottaviano Petrucci, a Venetian printer, went further still, adding a third printing for the addition of text and other details (like page numbers) – song anthologies he published in 1501, 1502 and 1504 are good examples of this.

Although Petrucci was relatively skilled in his art, and was the first to print polyphonic music using movable type, the process remained expensive, and thus cocooned from some of the obvious advantages that Gutenberg had achieved for printing alphabetical scripts. The breakthrough was to cast the stave lines onto the notes, thereby allowing a single impression to do all the work. Although Italy and France were generally the centres for sixteenth-century European music publishing, the technique seems likely to have been first developed in London by the barrister John Rastell, but by 1528 Pierre Attaingnant, a French printer based in Paris, was using the same approach – Attaingnant would be the first to use this method for significant print runs. Venetian printers followed. Paris and Venice were more natural centres for music printing than London as sixteenth-century Europe developed a distinct musical geography. Thus musical publishing would be centred on the German states, the cities of northern Italy and, to the west, on Paris and Lyons. Yet over the course of the sixteenth century it was increasingly Venice that came to dominate musical publishing in Europe, building on its longstanding Renaissance bookishness.

As early as 1468, Cardinal Bessarion had described Venice as a 'second Byzantium' thanks to its written scholarship. It was that year that Bessarion, a Roman Catholic bishop and one of the great scholars of the early Renaissance, donated around 750 of his own Latin and Greek codices to the city's library, the Biblioteca Marciana, as well as some 250 manuscripts and printed books. Bessarion's bequest pointed to the growing need for texts among Venice's scholars. The new combination of increased supply and demand in the world of books meant many readers now had leisure to pursue their own interests more exclusively. Venetian printers, inevitably, noticed this tendency towards specialization too, and began to follow suit.

The Scotto press in Venice enjoyed particular success in the second half of the sixteenth century. It printed on a range of topics, but its specialty was music, for which a market had now developed. The rise of printed music in Venice was especially good news for composers, since it enabled them to widen their circle of patrons; indeed, now patrons could simply pay for the printing of a book of music, which deepened the pool of possible patrons to include men who might not afford their own musicians at home. Composers could also do their own paper-based publicity and sell their music themselves. Venice's global trading networks made this all the more feasible and Venetian musical texts were sold abroad – in at least one case, as far abroad as Colombia.

As printed music became more commonplace, and as the type set for musical notation was used more frequently, so music was increasingly standardized, and its notation improved.[93] This tended to increase loyalty to the composer's original and to give the performer far less power to interpret the music himself (although this trend was fed by other movements within music too). But paper printing's greatest contribution to music came in its spread, as very different traditions became accessible to one another, enabling a close-knit community of European composers and musicians to emerge. As they fed off one another across the continent, composers profited both from more breadth in style and from greater competition among their peers.

The rise of Renaissance music also illustrates the impact that printed papers could enjoy in a specialist area, one with only a limited readership. Cost undoubtedly mattered less in music publishing than in, say, the publishing of religious tracts. Indeed, even single-impression musical scores were expensive enough (due to the complexity and variety of fount-types required) for printers to correct mistakes rather than simply use a new sheet; musical scores were the first printed texts on which, beginning in the fifteenth century, publishers used correcting fluid.[94]

The sudden flourishing of music in sixteenth- and seventeenth-century Europe was one of the most spectacular developments in which printing on paper would play a part. The Renaissance and Reformation were self-consciously derivative movements, looking back to various ancient sources (Hebrew, Greek and Roman). Composers in the sixteenth and seventeenth centuries did look back to the classical period for musical principles and for a philosophy of music, but they lacked ancient musical scores to serve as practical models. Renaissance music did, however, receive strong religious impulses. Indeed, the religious divide in Europe was one of the factors that lay behind the great miracle of European music in the age of printing.

Music now became a means both of theological affirmation and of competitive display. Luther himself played the lute and the flute and wrote many hymns in German for congregational singing, providing songbooks for his congregations. This marked a departure from the Latin traditions of the Mass, most significantly in the use of language and in its corporate emphasis. Luther harnessed music to his theological vision: the spiritual equality of all believers. He was mesmerized by the music of the great Renaissance composer Josquin des Pres. Above all, he believed in the remarkable power of music to affect the hearer inwardly:

> except for theology there is no art that could be put on the same level with music, since except for the [music] alone produces what otherwise only theology can do, namely, a calm and joyful disposition.[95]

When the Pope ordered a council to debate the Church's response to Lutheranism, music was a particular focus. Indeed, for all Luther's impact in his musical policy, the Roman Catholic Counter-Reformation, its start usually dated to the Council of Trent (which ran from 1545 to 1563), would be foundational for the rise of music in Europe through the rest of the sixteenth century, and into the seventeenth century too. The best composers would come from both sides of the divide, but the Counter-Reformation provided much of the institutional groundwork for what would follow. While early drafts on musical policy had provided some detail and strictures, the guidelines actually published by the Council in 1562 were remarkably light, pointing only to the need for purity and to a new focus on clarity and directness. Through providing direction in this way, the Council helped to encourage innovation, spectacle and beauty. Masses, the Council of Trent had stipulated, should be ordered, comprehensible and clear. The problems had been summarized by Bishop Cirillo Franco in a letter he wrote in 1549:

> I should like, in short, when a Mass is to be sung in church, that its music be framed according to the fundamental subject of the words, in harmonies and rhythms apt to move our affections to religion and piety, and likewise in Psalms, Hymns, and other praises that are offered to God . . .

> In our times they have put all their industry and effort into the composition of fugues, so that while one voice says, 'Sanctus' another says 'Sabaoth', still another 'Gloria Tua' with howling, bellowing and stammering, so that they more nearly resemble cats in January than flowers in May.[96]

The result of these rules was not a closing down of music but, instead, a continued flourishing. Moreover, the Council left the interpretation and enforcement of musical policy to be decided on locally; this looseness encouraged both variety and innovation. One of the Roman Catholic composers who shone most brightly under the Roman Church's new approach to music was Palestrina.

Giovanni Pierluigi was born in 1525 not far from Rome and since his death has been known by the name of his home town of Palestrina. His great strength lay in combining clarity with complexity, which was arguably the great challenge laid down by the Council of Trent as it sought to clarify music for the common ear. Palestrina was a master of polyphonic Renaissance music, perhaps most famously in his *Missa Papae Marcelli* (Mass for Pope Marcellus). For much of his life, he struggled financially and his personal letters reflect the difficulty composers had in making ends meet when they lacked wealthy patrons. In 1588 he wrote to Pope Sixtus V complaining that, although he had already published many works, he had many still waiting to be published but had been delayed by finances. He added that 'it requires a great expense, especially if the larger notes and letters are used, as church music clearly requires.' Publishing music, then, remained a major financial undertaking, and Palestrina would sometimes struggle in this area, despite his popularity with the Papacy, until his marriage to a rich widow late in life. He did publish hundreds of compositions throughout his life, however, and achieve fame before his death.

Despite the costs associated with printing music, other composers flourished in print as well. The works of Antonio de Cabezón, who died in 1566, were published by his son in 1578, with a print run of 1,200 copies. Moreover, printers experimented with different forms. Jacques Moderne, an Italian who settled in France and died in the 1560s, was one of the first printers in France to use the single-impression method. Moderne founded a music house in Lyons and appears to be the first printer to publish choir books with two-voice parts facing in opposite directions, so that people sitting either side of a table could sing together. Such innovation might seem surprising, given the relatively high cost of printing music. But, whereas many non-musical texts *could* be read and reread several times, musical scores were *never* intended to be read or used on a single occasion, but repeatedly. (Indeed, although initially aimed at specialists, published music would begin to enter the home in the sixteenth century too, and thousands of collections of both religious and secular songs would be published.[97]) The survival rate of Renaissance music books has been exceptionally low[98] but the reuse rate goes some way to explaining why.

Palestrina, however, was a particular publishing star of sixteenth-century music, favoured by popes and copied out by Johannes Parvus, scribe of the Papal Chapel, which relied entirely on manuscript sources. Indeed, evidence is lacking that printed texts were used by the Chapel even as late the Council of Trent. This unusual setup gave to Parvus extraordinary power and influence, reflecting a more hier-archical approach to book production, one driven not by market interest but by guidance from above. Parvus copied out a few of Palestrina's works, and the composer certainly enjoyed the patronage not only of princes but even of the Pope. Indeed, Palestrina's output was early proof that the Counter-Reformation had provided music with an opportunity, not a straitjacket.

Palestrina's success in his own lifetime meant that his works would travel and survive. Thus his influence was able to persist after his death, and greater, more prolific composers would learn from him, both Catholic and Protestant. Among them was the great Baroque composer Johann Sebastian Bach, born in 1685 in Eisenach some ninety-one years after Palestrina's death.

Bach, although a theological heir to Martin Luther, was especially influenced by Palestrina's *Missa Sine Nomine*, which he studied and performed while writing his own Mass in B Minor. As cantor and music director at St Thomas's in Leipzig, Bach had access to a good musical library; the St Thomas School library had especially rich collections of vocal polyphony from the fifteenth to seventeenth centuries.[99] Thus, despite the struggles Palestrina had endured, his music was widely available to his musical heirs, even more than a century after his death.

Unlike Palestrina, however, Bach won fame in his lifetime only as an organist, and his compositions remained largely unremarked in the wider musical world even at his death in 1750 and for several decades thereafter. Very few of his works were published in his own lifetime. One exception was *The Well-Tempered Clavier*, an initial compilation made in 1722 with a second set of preludes and fugues compiled in 1742. These, however, were not 'published' according to modern defi-nitions. Instead, they were merely circulated as manuscripts and not printed until 1801, fifty-one years after the composer's death. It might be tempting to see this as somehow a failure of print culture, but it can as easily be seen as an example of printing's success, since print offered the opportunity to revive a composer's work long after his death, perhaps even to grant him the worldwide and enduring fame he had lacked in life. Palestrina was merely revived by the nineteenth-century Romantics, whereas Bach's worldwide fame began with them.

Printed music had fed into Bach's own musical beginnings and development as well as providing an avenue for him to be rediscovered half a century after his death. His musical vision and depth owed perhaps very little to the medium of paper-and-print. But the success he achieved posthumously was very much a product of the print era. For it was only in the print era that composers could gain sudden and widespread fame decades or even centuries after their deaths, their works sitting on bookshelves, pianos and music stands in the homes of millions of admirers all around the world.

Printing on paper enabled composers and musicians to study, borrow, interact and compose music, and to publicize that music, far more easily. Public information campaigns were a crucial development in paper's age of print, and paper-based campaigns quickly became part of the urban landscape. Indeed, sometimes paper was viewed as the only way to address a public crisis; thus, from the 1570s, the problem of plague in England was addressed by printing out and distributing around public spaces procedures for citizens to follow and, soon enough, Bills of Mortality too. Under James I, plague orders would be published in the first year of his reign (1603) to advise citizens how and when to meet, what to avoid and what treatments and medicines might help them. (Even tobacco was publicly recommended.)

Once the 1603 outbreak had died down, however, paper returned to a more festive role, advertising pleasures rather than providing support during disasters. In London the 1603 outbreak had begun in Southwark, home to the city's great playhouses. As part of the emergency measures introduced by James that summer to stem the spread of the plague, playhouses had – as in some previous outbreaks – been closed. In 1604, however, life returned to normal in the capital and, instead of plague warnings, residents would see adverts for, among other amusements, performances at one or another of the London playhouses.

Playhouses were a particular feature of London's life and landscape at the turn of the sixteenth century. The master playwright of the age, William Shakespeare, was himself part-owner of one of the London playhouses, long since referred to as 'The Globe'. Shakespeare's plays enabled characters from across the social spectrum to appear on the same stage – kings and bishops, gravediggers and clowns. The stage therefore offered a snapshot of the full range of society to its viewers, enabling very different members of the audience to identify with at least one of the players. Standing in the pit in front of the stage were the 'groundlings', who paid just a penny for entry.

Shakespeare himself inhabited a paper culture and profited by it. Paper makes many appearances (and disappearances) in his plays, often, as in *Romeo and Juliet*, with great dramatic impact. But paper's most crucial role in Shakespeare's drama was the breadth of sources it delivered to the playwright himself. Indeed, it was during Shakespeare's own life that a new reading device appeared: the bookwheel. This remarkable contraption comprised a large drumwheel perhaps five feet tall, with shelves spaced across the inside of it. Designed in 1588 by an Italian military engineer, the device allowed a reader to have several books open in front of him at the same time and the books even remained at a constant angle as he turned the wheel.

It was a product very much of its time (although few were actually produced and used) since educated Renaissance readers liked cross-referencing and were therefore used to ranging between a selection of texts at the same time. Comparison was part-and-parcel of the Renaissance approach to learning. There is no particular reason to think that Shakespeare used a bookwheel himself, but the reading culture it served was his reading culture too, a culture that delighted in multiple sources, even for a single myth, event or metaphor. (Even a computer screen – with its multiple browser windows – cannot replicate this experience, in which various texts are fully visible before your eyes at the same time. The bookwheel experience was more like that of a latter-day financial trader, confronted by multiple screens.) This celebration of multiplicity had of course been enabled by the rapid increase in affordable books. Many of those titles made accessible in the Renaissance were texts from classical Greece and Rome and these became crucial source-texts for later Renaissance writers.

The culture of reading multiple books at a time, and thus of profiting from a range of sources in a single sitting, fed into Shakespeare's own work. He did not simply begin with a blank page; one of Shakespeare's attributes was his ability to pillage older texts for stories – and then improve on them – or simply plunder them for ideas, beliefs and metaphors. The range of this literary acquisitiveness was enormous; Shakespeare imported the world – and that included the world of books and learning – onto the stage. The plot of his 1599–1600 comedy *As You Like It*, for example, was taken from the Thomas Lodge prose story *Rosalynde, or Euphues' Golden Legacie*, which had appeared only in 1590. Even Shakespeare's title is taken from one of the phrases of Lodge's opening: *if you like it, so*. Yet in the play itself, the possible sources are so many as to make it impossible to trace line-by-line derivations with any certainty. And so it is in the speech, delivered by Jaques in Act II, scene 7, that hangs from one of Shakespeare's best-loved metaphors. It opens:

> All the world's a stage,
> And all the men and women merely players:
> They have their exits and their entrances;
> And one man in his time plays many parts,
> His acts being seven ages.

The metaphor itself, of course, was not Shakespeare's own invention. The image echoes *Policraticus*, an 1159 work by John of Salisbury which argues that a man's life on earth is a comedy in which everyone plays another man's role, while the choice of the number seven picks up on the biblical and medieval use of that number for religious and symbolic purposes. Renaissance influences are also strong. Erasmus's *In Praise of Folly* had described life as a play in which each actor waits for the director to motion him offstage, while Palingenius's book *Zodiacus Vitae*, first published in the 1530s, comes closer still, depicting the world itself as a stage. The 1570 publication of a world atlas in Antwerp entitled *Theatrum Orbis Terrarum*, often described as the first modern atlas and certainly recognized as remarkable in its day, may well have fed into the speech too – its name means 'Theatre of the World' (literally 'Theatre of the Sphere of the Earth'). It is a name that captures the excitement of the age of exploration, itself a rich seam in Shakespeare's plays. Or, looking back to classical antiquity, Shakespeare may have also been picking up on Ovid's *Metamorphoses* (a text he regularly echoed), in which the ancient Roman poet describes the four ages of man. Such was the range of sources made readily available by print culture at the end of the sixteenth century.

Moreover, while the accessibility of books helped Shakespeare to import the world onto the stage, they also helped him to export the stage into people's homes. Although the complete set would not be published until the first folio in 1623, the first of William Shakespeare's plays were published as early as the 1590s. The print runs were not large compared to the number of people seeing the plays themselves (usually no more than 1,500 were printed, while the Globe Theatre could hold 3,000 people for a single performance) but, by contemporary standards, a print run of 1,500 was not inconsiderable. *Hamlet*, in fact, saw a second quarto printing in 1604, just a year after the first. This time, however, the text ran to 4,056 lines instead of 2,221. That made it much longer – and therefore more expensive to stage – than the usual length of time allowed for plays, which was around two-and-a-half hours. But it did not matter, as this was a print run aimed not at players and their sponsors, nor even at playhouse audiences, but at readers.[100] Shakespeare's own name, unusually for playwrights of his time, could actually sell books, as publishers soon realized.

For the most part, however, Renaissance drama was not a publishing success, and many of its authors would only be published after their death. Meanwhile, Shakespeare's sonnets, with their intensely personal expressions of romantic love, also illustrate the shift from the private to the public realm that printing encouraged. Shakespeare's friend Francis Meres had praised the Bard's 'sugred sonnets' in 1598, but in the same breath mentioned their availability only to 'private friends' (as opposed to a reading public). Yet just a decade later, in 1609, *Shake-speares Sonnets* was published for a public readership, probably without proper consent, by Thomas Thorpe.

This personalized focus would gain much greater publishing traction in prose, however. Whereas poetry had always favoured memorization, brevity and performance, prose-writing posed a problem in the market because a typical reader could not afford to buy a parchment book (of considerable length) that he or she only planned to read once. But paper books, increasingly finding a standardized form (and thus lower price) during the sixteenth century in Europe, made such single-read purchases increasingly realistic. They also encouraged the practice of private, silent reading, where poetry had generally favoured performance or at least reading out aloud. The rise of printed prose would follow.

The novel's particular capacity for rapid and widespread sales was displayed in what is sometimes referred to as the first modern novel, *The Ingenious Gentleman Don Quixote de la Mancha*, or simply *Don Quixote*. Defining the novel is a dangerous pursuit, largely because it can take on so many shapes, echo so many different forms of literature and end up looking very unlike whatever definition you might apply to it. Henry James called it 'a loose and baggy monster'. But this characteristic, according to Cervantes scholar Anthony J. Cascardi, may in fact be what best defines the novel, namely, its 'ability to incorporate a seemingly limitless number of components and to assume an unpredictable variety of shapes'.[101] It is for this reason, writes Cascardi, that we can call *Don Quixote* the first novel. Cervantes's 1605 episodic account of the adventures of his picaresque, self-named hero Don Quixote and his loyal servant Sancho Panza was published in two separate volumes ten years apart (1605 and 1615). It quickly won an international readership. By 1620, *Don Quixote* had been translated and published in markets across Western Europe and parts of the Americas.

Don Quixote may be best described as a proto-novel, however, because it 'shows how the novel comes about when romantic idealism . . . collides with the real world'.[102] This makes sense of the timing of the novel's rise, which coincides with both the rise of modern science and a growing suspicion of classical authority. It is also a secular genre and closely tied to a new socio-economic phenomenon of the age of print: the middle class.

A novel is much more expensive to produce than a pamphlet or newspaper and therefore involves significant financial risk for the investor. That risk is only rewarded if a sufficiently large readership exists to purchase it, which means that the novel first requires a high number of non-specialist readers with enough disposable income to afford it – in other words, a bourgeoisie. But it requires more than that. It requires a shared language. Obviously that includes literacy in the same language, but it also includes a set of shared cultural experiences and understandings that can make the novel accessible to thousands of people who will never meet one another. This need for shared language, literacy and cultural identity across thousands of people is why the novel is often described as the literary form that is most closely tied to the nation – a hugely diverse set of people but one, nevertheless, with some degree of shared 'national' experience. France and England would experience its earliest flowering.

The breadth of the novel's readership marks an important moment in the history of paper's audiences, not simply because of its links to the bourgeoisie or to any kind of national consciousness, but because it crossed the gender gap more decisively than any serious literary form that had preceded it. The novel delivered itself directly to women as well as to men. More radically still, it was a form increasingly written by women, too. The Renaissance had provided a rationale for women's literacy. The Reformation had often gone further – Martin Luther had expressed his desire to see the vernacular Bible made accessible to women and, as a corollary, wrote that both boys' and girls' schools should be founded. But the novel would provide a serious avenue for female authors to write independently of men – and for a mixed audience.

In France, the emergence of female authors was aided by the development of printing (which took books into bourgeois homes), by the Reformation (since it encouraged personal Bible reading), especially in the south-west, and by the influence of Italy, with the spread of Renaissance humanist ideas (especially under King Francis I (1515–47)) to both men and women. Thus Francis I's sister Marguerite de Navarre's mix of learning, royal connections and Protestant sympathies led her to write on a wide range of subjects. She corresponded with an admiring Erasmus and wrote the *Heptaméron*, a collection of seventy-two stories, including one about doomed love, '*Amadour et Floride*', which has been described as a mini-novel.[103] Women writers received even greater encouragement in the seventeenth century, when Marie de Gournay published *The Equality of Men and Women* (1622), in which she argued for education for all, and *The Ladies' Grievance* (1626).

The literary salon, an Italian invention, would blossom in seventeenth- and eighteenth-century France as a centre for the discussion of books and ideas. This privileged institution especially favoured women, since it was believed women were better at conversation than men and more adept generally in social relations. Meanwhile, while the literary focus of seventeenth-century France was on the theatre, with its male playwrights and strict classical forms, women increasingly turned to narrative, which opened the way for more personal styles and subject matters. The novel was a natural form to choose, and one in particular stands out as symptomatic of the changed relationship women were seeking with the printed page.

La Princesse de Clèves was written anonymously and published in 1678, but its author is thought to be Madame de Lafayette, who had been born into a minor noble family in Paris in 1634. Often called the first modern French novel, the psychological tale was an incredibly bold story to tell in seventeenth-century France, not least because every character except the lead was based on someone at the royal court. It provoked widespread debate. Discussion centred around the principal character, Mademoiselle de Chartres, who marries the Prince de Clèves and then falls in love with another man. (As in some of the greatest English and French novels of the succeeding two centuries, what form the woman's name takes in the title of the work is indicative of her relative social independence. Like Madame Bovary but unlike Emma, the heroine in Lafayette's novel is named after her husband.) She avoids being unfaithful but informs her husband of her feelings. Before he dies (of a broken heart), he asks her not to pursue her true love. She then decides to reject the open advances of the man she loves, choosing instead to live in a convent, where ultimately she will die young: she is convinced that her true love would one day tire of her and forsake her for another woman. One of the elements that surprised readers was that the main character did not behave according to established social norms – this was, perhaps, all the more shocking coming from a woman. Lafayette had presented a very different kind of literary heroine to the ideal figure held up by French high society.

La Princesse de Clèves described a rarefied social world that most readers would never have seen, but the novel has since proved itself to be a far broader genre. Written in popular language, it was also the first European genre to treat the common people 'with unwavering seriousness'.[104] In England in particular, it began to evoke a range of characters from all walks of life. Moreover, it did not just borrow all kinds of characters; it borrowed all kinds of writing, too (although this reduced with the emergence of the realist novel in the mid-nineteenth century.) In this, it linked to a form of literature that appeared in a very different social context: the gospels. The New Testament gospels were themselves literary hybrids, plundering a wide variety of ancient and classical forms to tell their stories: history, apocalyptic literature, poetry, healing stories, parables and so on. Moreover, the gospels were similarly unusual in treating common people as serious human subjects, which was strange to the classical worldview. This, it has been argued, was where the idea that everyday life is precious found its origin.[105]

The novel's more democratic elements were among the reasons it was often looked down on by educated men of status. Many eighteenth-century commentators were not impressed by what they viewed as a populist and low-brow literary form, one suited to those who lacked a proper classical education. In other words, they saw it as suited to women. This initial disdain played into the hands of women authors and readers alike, freeing them to experiment with the genre relatively unhindered. Moreover, the novel's focus on the inner life only added to the sense that it was 'women's writing', since questions of the heart were widely viewed as women's business.

What followed was an unprecedented surge of women's writing, especially in England. If a classical education was not necessary in order to start writing for others, then this freed the novel from some of the old limitations of dramatic action, which often located the story in classical, royal, ecclesiastical or battlefield settings. Instead, it was now possible not merely to enter the drawing-room, dining-room, kitchen or marketplace but to use those locations as backdrops – focuses, even – for serious literature. The novel's particular emphasis on social relations, often through the vehicle of reported conversation, brought it into the everyday world, a world inhabited by those who had not received a formal education in history, the classics, theology or poetry (generally, women and the lower socio-economic classes) as well as those who had.

Inevitably, this meant not just female authors but female readers, too. As early as the seventeenth century, there were growing numbers of women readers in England's urban middle class, since technical books were increasingly written for just such a readership.[106] In physical form, the printed book was also portable, allowing it to be carried into the home. (Indeed, novels could be concealed, as they were by New England mill girls going into work in nineteenth-century America.[107]) The popularization of novels among women encouraged the domestication of reading; the fact that the novel was often dismissed as a more feminine literary form only helped to smooth its passage into female hands.

By England's Victorian era, the novel had become wildly popular. Thanks in part to the works of Jane Austen, the early nineteenth century saw the genre develop a respectable reputation in England. A publishing boom ensued. Between 1830 and 1900 some 40,000 to 50,000 novels were published in England, by close to 3,500 different authors.[108] This was a crucial period for publishing, too; no longer the work of printers and booksellers, by the nineteenth century publishing had become a distinct practice with distinct practitioners.

Of all printing's deliveries to the modern readership, the novel ranks as perhaps the most personal and individual of the mainstream products, even though it is seeking to appeal to an exceptionally wide readership. A novel is not simply an item you buy for yourself, like a newspaper, but an item you tend to keep and often develop a sense of ownership over. It does not dictate when you should read it (unlike the newspaper or work papers or the post). It has strong historical tendencies but it does not primarily seek to inform you about processes, events or even people in the world around you (as in, say, history or science). Instead, it asks you to internalize a story and to imagine it yourself – characters, usually locations and often much more. Inventive in its own right, it also seeks to feed invention in the reader's mind's eye.

Portable, affordable, accessible (at least in principle), individual, modern and middle class, the novel is one of paper's best-suited passengers. If the paper book were ever to approach its end, then I suspect the novel would be the last form to fall.

16

By the Cartload

Give me the liberty to know, to utter, and to argue freely according to conscience, above all liberties.

John Milton, *Areopagitica*, 1644

The city-states of fifteenth-century Italy transformed how Europe used the written word. Papermaking had begun on the peninsula in the thirteenth century, but it was the Renaissance which encouraged Europe's new technology to be put to far greater use by the ruling families of Italy's various city-states. There were many ways in which the power of writing was expanded at that time, building on paper's use in government, legal and administrative documents in the previous century, but its most extraordinary impact would be in politics. In the modern era, we still inhabit the world of bureaucratic government which those city-states pioneered. Yet while paper aided the centralization of politics, it also opened it up to wider scrutiny. As bureaucracy increased, so too did communication about politics. From this new culture of political correspondence a news media would emerge – one that still feels familiar today. What began in Italy would become a European (and then an American) movement, as a paper-based political conversation became increasingly public. There was nothing uniform about how this happened across the continent, but there was one moment of particularly radical breakthrough which rightly stands as a symbol for how politics – and information about politics – was gradually changing ownership. Paper, as we shall see, sat at the centre of the change.

In fifteenth-century Italy, however, one of the early drivers of the new writing culture was the rise of a fundamental Renaissance phe-nomenon – curiosity. Scholars wanted to study texts in detail, understand them and then remember them. Scientists hoped to do the same with nature. A thousand private obsessions were able to bloom among scholars at this time, thanks not to print, but to paper itself. Renaissance scholars in Italy already had access to good libraries, after all. What they now began to do – and in earnest – was to take notes.

Some of them tried to point to classical precedents for note-taking, often to the first-century AD polymath Pliny the Elder, who left almost 160 volumes of his personal notes to his nephew. If note-taking was common in the classical world, the evidence has (unsurprisingly, given the materials used) not survived. Medieval note-taking, which would have survived much better thanks to the durability of parchment, was presumably quite minimal, given how little is left today. The fifteenth century, however, saw a rapid shift, thanks to a fresh curiosity about a whole range of subjects. One man who encapsulated this new fascin-ation was Leonardo da Vinci. His notes stretch across some 6,500 pages – and those are only the pages that survive. Readers and schol-ars in fifteenth-century Italy, and later across Europe and North America, would carry small books around with them in which to take notes on whatever they learnt or noticed. In the century that followed, new words would appear in English to describe this innovative way of recording and understanding new learning: 'commonplacing' and 'excerpting'.[109]

Yet if private note-taking was a scholarly hobby for Renaissance men, keeping paper political records was a sine qua non for those working in fifteenth-century Italian politics. In 1448 the bishop of Modena, Giacomo Antonio della Torre, described the written output of the ambassadors of Italy's city-states as a 'world of paper'.[110] It was a despairing cry. Gutenberg's invention had made its recent appear-ance, but the papers della Torre was referring to were handwritten, the result of a crescendo of government documentation that had begun in the fourteenth century and reached a new peak in the fifteenth.

Behind this growth were increased political centralization and the rise of urban culture, with its more complex governmental demands, but equally significant was the higher profile of diplomatic relations between the Italian city-states after the Peace of Lodi in 1454 and the formation of the Italic League in 1455, which enabled peace between the northern city-states. Rulers now began to employ resident ambassadors in other city-states, and then further afield. Indeed, as early as 1497, the ambassador in England of the Sforzas (who ruled the Duchy of Milan) wrote back that the English king knew so much of Italian affairs 'that I could imagine myself in Rome'.

Such men would receive letters from their home ruler on a regular basis and could send daily handwritten reports back. Moreover, registers were kept of letters sent and so too were files of letters received, while written decrees were made for new appointments. One study found that in the fifteenth-century archive for the state of Gonzaga, of the 3,719 boxes, more than 1,600 were dedicated to diplomatic activity (and these were generally stuffed full). Likewise, the surviving letters of Lorenzo de Medici (1449–92), Italian statesman and effective ruler of the Florentine Republic, number at least eleven volumes and more than half are taken up with foreign policy. Although this new culture of documentation raised the importance of the Chancery or record office, its Chancellor and the process of archiving (especially as qualification expectations increased), what it failed to bring was a level of organization commensurate with the amount of paper bureaucracy now being produced. It would take centuries to impose some kind of order on the archives.[111]

What had begun in the 1450s after the Peace of Lodi intensified in the 1480s, as Florence and Venice normalized the exchange of ambassadors. Moreover, as power was centralized, it was decided that diplomatic correspondence should be the property of the state. Laws began to require the copying of diplomatic letters and other documents and the duplication of all diplomatic archives, since both the state and its diplomat kept approximately parallel archives of the ambassador's correspondence. Indeed, some ambassadors complained that they were unable to fulfil their duties because the written reporting responsibilities took up so much of their time.

One result of this need for news was the rise of the *avviso* ('notice', 'warning' or 'announcement'). Venice was producing its own *avvisi* in the sixteenth century as a way of informing (or potentially misleading) selected readers concerning its political, military and economic goings-on. Rome soon followed suit and these two cities began to produce regular news reports that were distributed elsewhere. These were key sources for ambassadors writing up their own political reports to send home. They were not quite the forebears of printed newspapers as they were very much based around Italian political intrigue, but they were nevertheless symptoms of a news culture beginning to emerge. This was especially true from the late sixteenth century, as *avvisi* began to attract a wider public readership and as (from the 1590s) Western Europe's postal service improved, centred on Germany. In the seventeenth century, Venice would provide much of Europe's news, distributing it to cities as far away as London, Paris and Frankfurt.

Commerce in sixteenth-century Europe also came to rely increasingly on correspondence between trade partners, as they sought to gather news of anything which could affect exchange rates, commodity prices, transportation costs and risks, new markets or wars and laws that might hamper trade. So it was that the practice of politics and commerce was increasingly oiled by written correspondence of news. Aided by cheap paper and the improved postal system, weekly reports became increasingly common and travelling merchants would share them with one another, reading letters out to their peers at inns or trade fairs.

If a news culture was now spreading across Europe, it was helped not only by the growing interest in foreign news but also by the phenomenon of printing. By the end of the sixteenth century, political events in France and the Netherlands were being followed by readers in Germany and England. The sieges of Paris (1590) and Rouen (1591–2) during the French Wars of Religion were described in English in London publications. The accessibility of such events to a public readership inevitably made that same public more engaged and concerned with the events themselves, a fact which would have enormous political consequences.[112]

Although some printed news-books and news-pamphlets were published in major European cities in the sixteenth century, notably in Germany, it was only in the seventeenth century that newspapers – sheets of two or four pages, initially with two columns, published at least weekly – began to appear. After decades of news pamphlets, *Relation aller Fürnemmen und gedenckwürdigen Historien* (Collection of all Distinguished and Commemorative News) was published in Germany in 1605. It is considered the first newspaper in the sense that it was the first regular, public publication that covered current events, but not in terms of layout, since it bore only a single column of text on the smaller, quarto pages.

Only in June 1618, when the Dutch *Courante uyt Italien, Duytsland, &c.* rolled off a press in Amsterdam, did the newspaper find its form – in double columns. Although printed on only one side, the Dutch newspaper was formed from a single folio sheet folded once (to form four sides) and opened at the fold. It was thus the first 'broadsheet'. The date and serial number both appeared only in 1619 and, from 1620, it appeared with text printed on both sides. At this point it can only be called a newspaper. An English version also appeared in 1620, published in Amsterdam, although this was a news-book in format and English news only broke out of its book form in the 1660s. A French newspaper appeared in 1631, initially called *Gazette de France*. 'Gazette' was one of the first new words coined as a newspaper title: the word derived from a Venetian monthly, first published in 1556, which cost a *gazetta*, a small local coin. And an inaugural American newspaper was published in Boston in 1690 – *Publick Occurrences Both Forreign and Domestick* – but after only one edition it was suppressed by colonial officials. The 1704 *Boston News-letter,* however, would last until the American Declaration of Independence in 1776.

In England, it was a Dutchman who was most instrumental in establishing the printing trade and, in the process, setting the precedent for which part of London would dominate the reporting of news. Although William Caxton had set up the first English printing press in Westminster in 1486 (importing papers from the Low Countries), it was Wynkyn de Worde who enjoyed the larger legacy. Indeed, some of the papers produced by England's first (and short-lived) paper mill, established by John Tate in Hertford in 1494, survive in the books de Worde printed.

When Wynkyn de Worde arrived in London at the end of the fifteenth century, he quickly fell on Fleet Street with its collection of monasteries – Whitefriars, Blackfriars and the Knights Templar – as the place to set up business. As the centre for monastic copying, Fleet

Street had become a legal highway too, since lawyers needed scriveners to write out their contracts. De Worde took over two houses next to The Sun pub in Fleet Street. He lived in one of them; in the other, he set up his printing press. What de Worde began soon attracted fellow printers to the area too, allowing a small hub to develop. Fleet Street would prove to be the crucible of England's revolution in mass communications, but in its early days de Worde's work dominated. In forty years, he produced in excess of 800 publications, more than a seventh of all pre-1557 printed works produced in England.

In 1513 the first recorded news-book, an account of the Battle of Flodden, was printed, though not by de Worde. It began:

> Hereafter ensue the trewe encounter or Batayle lately don betwene Englade and Scotlande.

But most of the great firsts in English printed media came in the seventeenth century, rather than the sixteenth. In part, the reasons were similar to those that spurred publication of the 1513 news-book, namely, peace and war in Europe. The beginning of the Thirty Years War in 1618 had implications for London traders doing business in Europe. Moreover, a 1586 ban remained in place on the reporting of English news. Thus the first titled English newspaper, published in 1621, was called *Corante: or Newes from Italy, Germany, Hungarie, Spaine and France*, and was translated from Dutch (but printed locally) and contained (as per the title) no news from England itself. English news would only become popular once the Star Chamber had been dissolved (in 1641) and England had begun its own civil war.

Perfect Diurnall (i.e. daily), the first authorized Parliamentary news-book, was published in 1642 and *Publick Advisor*, the first all-adverts paper, in 1657 – it included an advert for one of England's earliest coffeehouses. These were the first stirrings of popular newspapers and Fleet Street would, of course, become the centre for their production. Speed of communication also became increasingly important, a development reflected in some of the increasingly popular newspaper names, such as *Express* and *Dispatch*. Indeed, the English developed a hunger for the latest news and this hunger found a natural home in the coffeehouse.

The coffeehouse was peculiarly suited to the new culture of paper news. Less bawdy than the ale house or tavern, it provided a forum for debate about news and politics. From the second half of the seventeenth century, coffeehouses appeared in both England and France, and English coffeehouses in particular allowed their clientele to discuss both news and opinion, like popular informal parliaments. In coffeehouses, people could associate with each other freely, express themselves freely and enjoy the growing freedom of the press. As early as 1712, the *Bristol Mercury* complained of their popularity:

> About 1695 the press was again set to work, and such a furious itch of novelty has ever since been the epidemical distemper, that it has proved fatal to many families, the meanest of shopkeepers and handicrafts spending whole days in coffee-houses to hear news and talk politics, whilst their wives and children wanted bread at home, and, their business being neglected, they were themselves thrust into gaols or forced to take sanctuary in the army.[113]

Visitors to London were in awe of the impact of this new obsession; César de Saussure, a Swiss diarist visiting London in the 1720s, wrote that the coffeehouses were crowded and smoky and filled with the reading of news and politics. This fever of interest in the events of the day helped to cement the place of the newspaper at the heart of urban life. And it was Wynkyn de Worde's Fleet Street base, where he had first set up his presses, which would remain the geographical heart of the British press until the media exodus of recent decades.

In the 1670s the words '*Zeitung*' ('Times' in English), 'newspaper' and 'journal' all entered the printing lexicon. The first 'magazine' appeared in 1731, the word taken from the Arabic *makhazin*, its title a reference to a military storehouse of various materials since its content was so varied. Yet only improved literacy could deliver anything like a modern newspaper; daily publication finally began in the early eighteenth century in England and mass distribution, in America, in the middle of the nineteenth. Access to news does not have to imply any power to shape it. Yet without proper access to information, it is impossible to engage meaningfully with political issues. Moreover, access does at least imply engagement, at whatever distance. In this lay enormous opportunity for the readers – and for the newspapers, too.

*

By the end of the sixteenth century the United Provinces of the Netherlands were rivalling Venice as the intellectual print hub of Europe. In 1620 Antwerp would lie at the heart of two international postal systems carrying mail to cities in England, France, Spain, Portugal, Germany and Italy. Moreover, the Dutch Revolt had created an opportunity for print. The seventeen Dutch provinces had risen up against Philip II, ruler of Spain and of the Spanish Netherlands, beginning in 1568*¹and many in the Catholic south had emigrated north, among them printers and booksellers.

From its starting base in Antwerp, which dominated Dutch book production in the sixteenth century, the Dutch publishing industry was able to prosper in cities across the United Provinces, thanks to favourable economic conditions (low interest rates and high availability of capital) and an excellent transport system for trade. Moreover, the devolved political set-up that the Provinces enjoyed meant that restrictions on publishing were less onerous than elsewhere in Europe, while copyright protection was unusually strong. Debate and publishing thrived in the Netherlands together, and many controversial foreign authors found their publisher in the Netherlands' liberal market. The Dutch presses were also helped by the greatest technological improvement made in papermaking for a millennium: the Hollander. Created in 1680 in the Netherlands, the Hollander used blades rather than merely wooden stamps to macerate paper pulp – in one hour it could produce as much pulp as the old hammer mill had taken a day or more to pound up. The paper fibres were weaker, but production efficiency was transformed.

By the mid-1700s Amsterdam, The Hague, Leiden, Utrecht and Groningen all had their own *couranten* (newspapers). There were even French-language newspapers produced for the large numbers of French Huguenots who had come to the United Provinces in the seventeenth century.¹¹⁴ Although universal education and literacy were not widespread political aims until the nineteenth century, the eighteenth-century public in Western Europe was increasingly a reading public rather than merely a listening public, and the output of Europe's presses both drove this change and responded to it. Nowhere had the change been more marked than in the Netherlands where, by the end of the

---

* Its start is sometimes dated to 1566.

seventeenth century, illiteracy was the lowest for any European country. One study made in the 1960s looked at how many brides and grooms in Amsterdam could sign their names in 1630, 1680 and 1780. Among men, the percentage rose from 57 to 70 to 85 per cent across the period, and the percentage for brides rose from 32 to 44 to 64 per cent.[115]

The United Provinces were increasingly the centre of the printing trade in Europe, with the technology, money, networks, literacy levels and political openness to mass-produce even dangerous political texts, and it was the free thought and printing presses of the Netherlands which would provide an especially important printing launch pad for revolution in France. Although Dutch readers did not enjoy a press free of state controls, they did inhabit a divided state, with different groups controlling different regions. This splintering of power gave writers and printers practical freedom, since they could generally find a region where their writings could be safely printed. Spurred by French Calvinism, Amsterdam became a focus for the latest Protestant writing. Dutch Protestants in the sixteenth century had taken many of their arguments from French Calvinist writings and their Antwerp presses had followed events in France closely. But French Protestant writings peaked in the 1560s and, by the 1580s, religious writing in France itself was almost entirely Roman Catholic.

Pierre Bayle, a French Calvinist philosopher, fled first to Geneva, then returned to France in 1675, then fled again, this time to the Netherlands in 1681, where he would remain until his death in 1706. From the city of Rotterdam he was able to publish a number of works, including his literary journal *Pensées sur la république des lettres* (*Thoughts on the Republic of Letters*) between 1684 and 1687. This was followed by his *Philosophy Commentary*, also in French, which argued for religious toleration. Geneva and Rotterdam were able to offer Bayle not only freedom from the risk of persecution, but also freedom to print his views – and even the prospect of having them circulated back in France.

More telling still is the case of the *Index of Prohibited Books*, a list compiled by the Roman Catholic Church to ensure that Catholics across Europe would not print or read certain titles. (There would be different reactions to the list among Catholic governments, and sometimes a government would use its own list instead, but often the same titles were included.) Thus, although a few rich readers could access them illegally, official printers of these works could be taken to court, which helped to keep such titles from reaching too large an audience. In the Netherlands, however, where such restrictions could be evaded, one Dutch printer actually used the index for his own publishing programme, since he was convinced these censored titles would sell especially well.[116] Thus the Netherlands provided print spaces for disaffected European thinkers to make themselves heard – even across the continent – and such freedom was especially important in allowing the exchange of ideas about French politics and society (and the ideals that should underpin them). Amsterdam in particular – open, religiously tolerant and, at least in principle, devoted to free trade – welcomed Spanish Jews, French Huguenots, Flemish merchants and religious refugees from Spain's remaining southern provinces. The city quickly became a haven for goods and ideas: books were the inevitable effluent.

One innovator who used Amsterdam to publish his works was Abraham-Nicolas Amelot de la Houssaye (1634–1706), a historian employed as an ambassador by the French court. As a humanist editor and publicist, de la Houssaye ensured the publication of several major works of history and rhetoric of the period. He flouted publishing conventions by selling the text of secret treaties and embassy letters but also by publishing his own political works, works that laid bare the workings of the French state. De la Houssaye made a very bold political statement by publishing his books with notes, since textual criticism was a form of political criticism. While he did not push for revolution, the very act of revealing the workings of the state and providing commentary on them opened French politics up to outside opinions.

De la Houssaye had at least fifty-nine editions of his works published in French before his death in 1706, and almost the same number again were published during the eighteenth century. His most significant publication was a translation of *The Prince* by Machiavelli, which he produced in Amsterdam with accompanying notes. De la Houssaye's translation was re-edited in Amsterdam at least five times during his life and more than fifteen times during 1700s. First published in 1532, Niccolò Machiavelli's *The Prince* describes how to hold on to a state, and the overarching message is that the ends justify the means. Although some critics now believe the work was intended as a satire rather than a recipe for governance, it was taken at face value in its time, as it was much later by both Napoleon and Stalin – Napoleon made notes on the book and Stalin annotated his own copy.[117] De la Houssaye, who defended Machiavelli, had chosen to read (and then to write on) history critically and in this lay the seeds of political reform and change.

Works such as these were smuggled into France as a means of creating an underground Republic of Letters, an unofficial political and cultural debate free of government censorship. In this such reformers were, in fact, successful. Yet they were also, of course, involved in a movement that would prove to be far bigger than they could have imagined – not just for France, but for the world. After all, you do not have to view liberal democracy as history's final triumph to appreciate the symbolic power of the French Revolution for the age that we live in now. The press was not only involved in preparing the way. It became a particular focus of the revolutionaries' hopes for a new and freer society. Far more importantly, however, 1789 spelt out for a watching world what kind of paper culture should be the ideal. And that answer, despite how the French Republic developed in its first few decades, was the culture of the free press.

The press in seventeenth-century France was essentially a club of printers and booksellers who set their own rules. In 1686, Louis XIV had fixed at thirty-six the number of printers in Paris, the dominant centre of French publishing. Eighteenth-century Paris inherited this narrow system – a new member could only gain access to the group of thirty-six if an existing member died. Except for widows of those who had been legally in the trade, women were forbidden from working in printing, publishing or bookselling. Moreover, the printers were plugged into a large network of censors, police and others who enjoyed royal patronage. Within this structure, writers themselves existed as private individuals who required the legitimization of an absolutist state.

The book trade was monitored by three separate bodies; the most important, the Administration of the Book Trade, operated a system of 'privileges' that constituted a kind of permit to publish and proto-copyright. Only in 1777 did the law change to allow authors to publish and sell their own works. In essence, it was a system that derived from the belief that all knowledge came from God and was mediated through the king's licence. It was a publishing culture that favoured traditional views on politics and religion, and the king's licensing system ensured that this conservatism could not be legitimately challenged.

There was, however, another book trade in operation in eighteenth-century France. Since the French state set such onerous conditions on the publication of books, printing houses began to appear close to France's borders, often in Switzerland or the United Provinces. In this way, illicit books could easily be smuggled in, their path smoothed by bribing customs officials or simply concealing them among other goods. The city of Rouen became an especially important stop on the smugglers' trail as Dutch Protestant printing houses supplied Enlightenment writings to the French market. Illegal and underground, this alternative book trade viewed the author as the sole source of the work and sought to promote the ideas of great Enlightenment thinkers like Voltaire, Jean-Jacques Rousseau, Denis Diderot and Honoré Mirabeau. Thus Voltaire's *Oeuvres* (Works) was published in Kehl in Germany, and Rousseau's *Oeuvres* was published in Geneva. (His *Social Contract*, written in 1762, was a product of the Dutch publisher Rey. Rey himself was from Geneva, the son of French Huguenots, and never became fluent in Dutch, which is suggestive of how the book market was crossing borders.) Diderot's landmark encyclopaedia, initially published legally in 1750–55 in twenty-eight volumes, would later be published without official patronage in Geneva and Neuchâtel. (It was expanded to thirty-five volumes and had sold around 25,000 copies by 1789, half of them in France.) And while some of Mirabeau's pamphlets were published in Avignon, others were published in the Low Countries, especially in Amsterdam, as well as in Lyons, the Paris suburbs and Paris's Palais Royal district, which enjoyed relative immunity from the king's police. Thus Enlightenment culture was effectively diffused in Parisian society by a network of smugglers, dealers, publishers and printers inside and outside France itself.

Some printers and publishers worked on both sides of the law, and might even hear their illicit titles denounced from pulpits – or witness them burnt. Yet sermons could not stem the influx of printed works and, increasingly, radical books on politics and sex began to enter the country too. Aiding these new arrivals was a more general confusion over literature. New genres were appearing and nobody seemed able to manage and rationalize the book market. For example, French booksellers usually had a section called 'Philosophical Books', but its shelves might include both works by authors like Voltaire and more erotic works like 'Philosophy in the bedroom'. Often it seemed as though freedom and sexual licence were linked and even the great French Enlightenment authors wrote their own erotic works. Honoré Mirabeau, a leading light in Revolutionary thought, wrote both bold political tracts and pornography.[118]

Although these books were not directly urging France towards revolution, they were answering the reading needs of a public curious about politics, sex and news. The ability to read about and discuss hitherto censored issues signalled a change: literature was beginning to forge public opinion in new ways as members of the reading public were reading the same books, including the same philosophical works. Literature was attaching itself to a freer market. Vigorous printed criticism of the monarchy simply grew in audacity and volume through the eighteenth century until the king had lost legitimacy in the eyes of his reading subjects. The simple fact of making literature and history more accessible to the reading public turned them into something mediated to the individual through print, rather than mediated through a monarch and his administration. That carried in it the potential for socially conservative readings to be increasingly questioned.

The monarchy itself was not entirely opposed to some opening up of the publishing industry, and the man charged with regulating the country's book trade, Guillaume-Chrétien de Lamoignon de Malesherbes, was not just reasonable but surprisingly liberal in his own attitudes, encouraging the initial publication of Diderot's *Encyclopédie* and protesting against moves to centralize power. In 1788, a royal order called on 'educated persons' to articulate and publish their opinions on procedures for convening the Estates General, France's toothless legislative assembly. The Paris parliament even legitimized the concept of a free press in the same year. But it was events in 1789 which would carry the greatest weight.

Early in 1789 the Administration of the Book Trade faced a problem: the king had called the Estates General and asked it to discuss the trade, but the Estates General still expected the Administration to maintain its historic role. This meant it should be prescribing and policing who had the right to publish materials about the Estates General. Pirate works, however, were increasingly flooding Paris. Despite official pronouncements, print inspectors in towns and cities across France reported back that illicit publications had proliferated and were beyond their control. Moreover, the central administration struggled to provide clarity over what was permitted; after all, it was waiting on the recommendations of the Estates General – and the king's reaction to them. (The king himself refused to give guidelines until the Estates General had reported back to him.) The result was a de facto suspension of censorship in many cities across the country. Indeed, in Paris, three major national newspapers were given freedom to print anything 'prudent' about the Estates General.

A few months of unclear publishing policy from the monarch may not sound disastrous, but in effect the crown was foregoing its right to interpret political events. As the Administration of the Book Trade foundered, a number of regional inspectors absconded, stopped writing their reports or simply admitted defeat. The centralized administration of France's book market was collapsing and, as it did so, power was passing from the monarch to the market.

On 9 July 1789 the National Assembly was founded as a 'people's' successor to the old Estates General. Its first moves were to formally abolish feudalism, to sweep away a range of royal, noble and clerical privileges, and to publish *The Declaration of the Rights of Man and of the Citizen*. Article 11 stated:

> The free communication of ideas and opinions is one of the most precious of the rights of man. Every citizen may, accordingly, speak, write, and print with freedom, but shall be responsible for such abuses of this freedom as shall be defined by law.

In the story of paper, these two sentences marked a watershed: a talisman for the defence of free speech, free pens and a free press. Their impact on press culture in France was immediate.

In 1789, printing in France quite simply exploded. Royal patronage of print culture was now gone, meaning that print itself had been both freed and empowered. By December royal officials were complaining that everyone seemed to want to open a print shop – even in villages. But the centre of activity was Paris. The revolutionaries had hoped that the free press would deliver what they believed France most needed – a republic of Enlightenment readers, in which the works of Voltaire and Diderot were common currency.

What actually followed was a proliferation of print in the very city that had been at the heart of absolutist France: Paris. Once the freedom of the press had been proclaimed, the illicit publishing world of Enlightenment France suddenly stepped out into the open. Some printers appeared from the suburbs or from basements or from prisons. But it was not only local, hidden printers who began to appear. In 1789 and 1790, printing presses, loaded onto carts, began to converge on Paris from the cities which had once been among its chief suppliers of revolutionary works. In 1790 Beaumarchais announced that he would move his print works from Kehl in Germany to Paris. The *Encyclopédie*, which had for years been published outside Paris, was now peeled from the printing press in the capital itself. From 1789 Rousseau's *Oeuvres*, hitherto printed in Geneva, now announced on the title page 'Paris' as its place of publication. The presses of France's Enlightenment had come out of the shadows and out of exile to be wheeled into the capital, the city which now became the unquestioned centre of Enlightenment thinking and publishing.

The most dramatic moment in this transformation came immediately after the Revolution. Thus, while 226 legal printers, publishers and booksellers were in operation in Paris in 1788, in the decade following 1789 the total rose to 1,224 establishments. That number would drop a little in the early Napoleonic era, but around half of the print shops which survived in 1811 Paris had been founded in 1789–93. Likewise, in 1788 only 4 journals were published in Paris, but in 1789 that total rose to 184, and to 335 in 1790.[119] Paris was already the intellectual capital of the world, but from 1789 its print culture began to catch up.

Nevertheless, although Paris would retain its intellectual pre-eminence, the initial print culture expansion that followed 1789 could not maintain its momentum. Instead, with printing restrictions removed, publishers saw prices drop wildly and printed pamphlets and ephemera proliferate. The print revolution of 1789 did not see the established printers of Paris churning out the great Enlightenment works for a new nation of philosopher-readers. Instead, it started a furnace of publishing with new, small-scale

LIBERTÉ DE LA PRESSE

16 The press freedom that followed the 1789 Revolution looked like mayhem to many onlookers, and appeared to push publishing towards sordid subjects. The liberated people of France were not, as some had hoped, addicted only to Voltaire, Rousseau and Montesquieu. By 1797, when this image was printed, many were in despair at the apparent social havoc caused by unchecked printing. The process itself continues unruffled behind the crowd: choosing letter-types (right), inking the press and screwing down the platen (centre), peeling off the pages (left) and leaving them hung out to dry (top). (Copyright © Bibliothèque Nationale de France)

printers producing revolutionary political ephemera. Catalogues and surveys from the period report that Paris had 47 printers in 1789, 200 in 1790 and 223 by 1799.[120] It was not the book which won in 1789, but the pamphlet and periodical – not high learning, but democratic access.

The new French print culture did not last very long. It bankrupted many printers through the late 1790s and into the 1800s, provided no rights for authors and spawned a kind of print mayhem that was as destructive as it was creative. In response, the authorities would first begin to protect the market and its authors, and then start to control the output. However, instead of using direction and patronage as the means of control (as had been done before 1789), they employed censorship and surveillance. Afraid that novels and other forms of low literature were too popular, the state increasingly provided funding for what it viewed as the best kinds of literature, leaving market forces to provide the rest. As a result, the book market contracted in the early nineteenth century. It also became more conservative in the titles it produced. Where the logic of the 1789 Declaration had implied a republican book trade, Napoleonic France would instead run it along familial lines. This was most demonstrably the case with the extension of copyright to twenty years after an author's death – *and* the death of his or her spouse. This not only made the book a more valuable product for authors, it also confirmed that ownership of the book would not immediately pass to the people, as if the author were simply a public servant. It was, in these respects, a disappointing conclusion to a remarkable opportunity.

And yet the government still had to defend its actions – and to lend its support to the idea of press freedom that the Declaration had articulated. This was more important for the uses of paper than the new technologies that enabled it to be produced at such improved speed and cost in the industrial age. Lithography, invented in Germany at the end of the eighteenth century, enabled joined-up writing to be printed. The Fourdrinier papermaking machine, invented in France, obtained its first patent in 1799; it became the basis for papermaking machines worldwide (and still is) as it produced one continuous sheet of paper, and enabled a much greater range of thicknesses and lengths. And the steam press, first set in motion by its German inventor in

London in 1810, meant the printing press no longer needed to be worked by hand; in 1814, an edition of *The Times* rolled off one of Friedrich Gottlob Koenig's presses for the first time, signalling a new era for newspapers. Paper's price also continued to drop – by half in the Netherlands in the late nineteenth century and to negligible levels thereafter. Today, of course, it is among the cheapest everyday products you can find.

Technology has been the partner of paper's universalization, aiding its spread across society as well as its increased use by any given individual. The modern pursuit of universal literacy, a campaign undertaken on paper (until very recently), is testament to the high regard that words on paper enjoy and the sense that without access to them, any individual is unfairly disadvantaged. Paper's ubiquity in the modern age hardly needs spelling out – until the digital age, it had no competitors.

Paper's ability to make information widely accessible, and even affordable enough to buy for your own bookshelf, has not, however, achieved what 1789 appeared to promise. It is reflective of paper's particular qualities that the media should be referred to as 'the press'; newspapers and magazines are cheap, rapid-production, rapid-distribution, light and portable items. Thus they are embodiments of paper's peculiar strengths. The problem is that, on a global level, they have yet to come close to winning the battle that 1789 articulated for them. Of course, paper's partners were legion in the decades that followed 1789, from bureaucracy to private letters to novels to tickets to ideas. But at this point the print movements of Europe began to fan out in hundreds of different directions and they did so under a new ideal of pursuing press freedom. Although there were some significant legal changes prior to 1789, notably the first British copyright act in 1710 (one of the first laws to recognize the rights of authors), in most European countries strong cultures of censorship and of preferential treatment for certain groups would not begin to wind down until the end of the eighteenth century. It was the French Revolution which was most effective in spreading such liberal ideas across the continent (despite what would then occur in France itself), ideas which promoted the freedom of the press and therefore led to better copyright protection as well –

the author would win more freedom in what he or she said and more rights over publication, too.

The *Declaration of the Rights of Man and of the Citizen*, with its clause defending the freedom of the press, was adopted on 26 August 1789. It was rapidly translated across Europe and became a talisman for the legally enshrined defence of free speech and the free press. (So, too, in England, did Thomas Paine's defence of the French Revolution, in his 1791 work *The Rights of Man*.)

In America, meanwhile, a very different print atmosphere preceded revolution and the establishment of founding principles. (After a first mill built in Mexico in 1575, the next paper mill to appear in the Americas was opened in Philadelphia in 1690.) Before the United States Constitution was ratified, a debate-in-print took place over what that Constitution should contain. This debate was rolled out across the country, symptomatic of a desire to use reasoned argument as the foundation for the new America and to avoid privileged influence where possible. Nowhere was this more clearly demonstrated than in the publication of *The Federalist*, written in 1787–8 by Alexander Hamilton, James Madison and John Jay, future Founding Fathers and already well-known across America. *The Federalist* contained a series of essays arguing for the ratification of the US Constitution. In the first essay, Hamilton wrote:

> It seems to have been reserved to the people of this country, by their conduct and example, to decide the important question, whether societies of men are really capable or not, of establishing good government from reflection and choice, or whether they are forever destined to depend, for their political constitutions, on accident and force.

The authors, however, omitted to put their names to the work, instead signing it 'Publius' in honour of a Roman consul. Nor was their decision unique; anonymity was a widespread practice in the debate over the Constitution and, where used, it helped to create a sense of the debate belonging not only to the great men writing publicly on it, but to the whole nation. The logic was clear: reason alone should receive preferential treatment in the debate over America's future.

Public debate certainly featured in the run-up to the vote. In America, as in Britain, print had helped to forge the concept of public opinion. Unlike in Britain, however, with its established political centre in London, America lacked an obvious urban political hub and focus – this absence delivered to the press even greater power than it enjoyed in England.[121] Indeed, when the Constitution was published so as to launch a national debate, it was reckoned to have appeared in every newspaper in the country within six weeks. Reports appeared in the press that all ages were reading and discussing the Constitution. Indeed, a letter from Salem County which appeared in the *Massachusetts Gazette* recounted how:

> Nothing is talked of here, either in public or private, but the new Constitution. All read, and almost all approve of it. Indeed it requires only to be read, with attention and without prejudice, to be approved of.[122]

Debate on the page had characterized its passage when the Constitution was passed by the Constitutional Convention in 1787. That same desire to debate freely and openly characterized its content too, of course, as it included a clause giving authors exclusive copyright over their writings. Moreover, the American Bill of Rights, published in 1791, included the First Amendment to the Constitution, which stoutly defends, among other freedoms, the freedom of the press.

> Congress shall make no law respecting an establishment of religion, or prohibiting the free exercise thereof; or abridging the freedom of speech, or of the press; or the right of the people peaceably to assemble, and to petition the Government for a redress of grievances . . .

These two statements of press freedom, American and French, just two years apart, have endured in their home countries and been replicated abroad – with varying levels of wholeheartedness and success – ever since. The passage of the constitutions in France and the United States mark a decisive moment in the history of the surface on which they first appeared. The ideals they expressed pointed to a future in which the press would no longer be the tool of government but, instead, would be allowed greater freedom than the state itself. The press would not only be used to question and undermine governments and their leaders – it would be used to do so legally, and even expected to do so. In the aftermath of the 1780s and 1790s, the printing press would discover a new power, one which has obliged many European governments to work to win the press's support rather than simply to expect or demand it.

Yet even within Europe, and certainly across the world, that shift in the role of the printing press, which 1789 had pushed for and publicized, is still under way, its completion far from certain. However distant 1789 may feel, Article 11 of the *Declaration of the Rights of Man and of the Citizen* remains a widely felt hope rather than a global reality. Until the free press can be proclaimed victorious, we remain very much in the shadow of an ideal that finds its most historic expression in the French Constitution and the First Amendment to the American Constitution. Print was discovering new owners and new protections, the debates it carried now reckoned to be not subject to the government, but above it.

There were a few immediate handovers that spelt out this new form of ownership more clearly. Thus, in 1789, France's Constituent Assembly transferred ownership of all religious houses to the state; much of their library stock was taken in order to fill the shelves of new public libraries. The ownership of knowledge was thus delivered from the Church to the reading public.

That reading public may be paper's final destination and the logic of progressive modernism certainly points that way. In the nineteenth century, the uses of paper would become so varied and the ideas it carried so numerous that it could no longer simply act as the plaything of one or two major calls-to-arms. Instead, it would become the surface of societies across the continent.

Many of those societies were freer than ever before. Of course, printed books and newspapers are remarkably useful to totalitarian states as well, enabling the mass education, co-option and deception of entire nations. Yet even such public campaigns of thought dictation imply the value of universal literacy and of access to texts. That in itself accords a far more active role in politics for ordinary citizens than has been the case through most of history. It also tends to provide certain opportunities for those who find the official creeds hard to swallow. You are hard put to find a country in which the political elite does not at least strongly influence the press. Nevertheless, no government can entirely control it; you cannot remotely hack into samizdat magazines, banned books and underground newspapers, and those forms can usually find their way into the hands of the dissatisfied, disenfranchised and disempowered in the end.

The freedom to write and to criticize texts, a freedom that paper released, had given birth to a new Europe. Paper continued to be its surface but the continent had already been made in its image. It is these two facts – the new ideal of free speech that paper helped to deliver and the fact that the free press has not yet won its global quest – which give to the years 1789 and 1791 such symbolic power in the story of paper.

That story has, as yet, no end-date, but in the ideals of 1789–91 it discovered an identity that it retains to this day. It is an identity bound up not with its parenthood but with its heirs: the reading public. Paper today is so ubiquitous, affordable and compact that it is hard to establish an absolute monopoly on what appears on its surface. Given the twin human desires of enjoying personal freedom and controlling the freedom of others, this is an extraordinary strength. Paper has, thankfully, proven itself to be beyond complete control.

Epilogue
Fading Trails

17 Final phase: pressing newly made paper dry in central China. The photo was taken by Father Leone Nani, an Italian missionary active in China from 1904 to 1914.

In 1840 British colonists and their Maori counterparts signed the Treaty of Waitangi by the eastern shore of New Zealand's North Island. Paper was now making itself felt in what the British called 'the Antipodes' (or, more colloquially, 'the back of beyond'). The name was not entirely without justification; the lands of New Zealand were among the last places on earth to be settled by humans. As for humans, so it was for paper.

The Treaty of Waitangi is a rarity in imperial history. Instead of simply imposing British rule on the three islands of Aotearoa, Te Waipounamu and Rakiura, a bilingual document was written up for both sides to sign. The Maori had encountered words on a paper page only when Europeans arrived on the islands; in approximately twenty years, between around 1820 and 1840, the culture of the islands had moved from oral to manuscript literacy to printing.[123] In 1836, the Cornish Protestant missionary William Colenso had printed the Declaration of the Independence of New Zealand and, in 1838, the first Maori New Testament. Now paper was being used to create a lasting political settlement that both sides could agree on. (Colenso also printed the Maori version of the Treaty.) For all its flaws, the Treaty of Waitangi was a remarkable attempt to find a political settlement that both sides could agree to, without shedding blood. It would be taken to key settlements around the islands and signed by more than 500 Maori leaders.

The treaty's greatest flaws lay in its modernist assumptions, which belonged to the age of literacy. First, the Maori translation was of variable quality, and in its most important element it was amateur. At best, it assumed equivalence between English and Maori words that described very different understandings of governance and rulership. The result was that, while the English version proclaimed Britain to be sovereign over New Zealand, the Maori translation suggested ultimate sovereignty was retained by the native Polynesians. New Zealand today celebrates and claims its pre-European heritage far more than its post-imperial peers. Despite this, the poor translation of the Treaty of Waitangi continues to be a stumbling-block in resolving Maori–*Pakeha* (Maori–European) differences.

The treaty also raises a second, more difficult question. If the pre-European tribes of New Zealand were oral cultures, why should an agreement written out on paper bear more weight than a spoken treaty? Paper documents may be how the *Pakeha* operated, but the *Pakeha* had arrived as outsiders in lands inhabited by Polynesian tribes. Should the future of those lands be decided with a European import: words on paper? Was this disembodied and silent form of speech really more valuable than the spoken word?

Paper's arrival on the Polynesian Islands should have marked its final triumph. From China, papermaking had travelled into South-East Asia, Central Asia and the lands of Islam. Islam had ensured the widespread adoption of papermaking culture in the Indian subcontinent – although paper had been used there centuries earlier – and had passed it on to Europe. The Spanish (and their European successors) had then taken papermaking to the Americas. Islam, Christianity and global trade had been just some of the influences which normalized papermaking on Africa's coastline. The islands of the South Pacific were therefore among the final destination points on paper's journey around the world.

Yet today the Treaty of Waitangi, which should have been one of the paper text's great triumphs, appears instead to proclaim its drawbacks. It assumes that literacy is the privileged means of communication: not merely that it is more definitive, but that it should be, too. The history of New Zealand, however, suggests otherwise. Or, rather, the two histories of New Zealand – oral and written – suggest that the truth is less clear. The immediate popularity of the Bible in Maori culture, following its introduction by literate Europeans, illustrates the differences between these two histories. The Bible itself is full of the written versions of Hebrew oral histories and the Maori absorbed many of these stories because their structures were often more familiar to the Maori setting than to literate Europeans. This reflected the two ways of telling New Zealand's history. In the Rongowhakaata tribe, for example, it was the women who composed the *oriori*, lullabies by which children were taught their family and tribal history. This was very different to the modernist (and male-dominated) means of communicating history; it used different narrative structures too. Paper has its own particularities and limitations.

More recently paper text has received more direct threats to its primacy. The first came with the invention of radio, which allowed you to listen to distant voices as they spoke, as you sat on your sofa at home. However, it could not, unlike printed books, offer you choice over when you listened to your favourite programmes. Images, on the other hand (photographs, and moving images in particular) might have made printed text increasingly obsolete. However, photography at least (its output increasingly decoupled from paper) has proved to be a less effective form of communication than the printed word. Photography and video each suffer from a major drawback, both of them outlined by Susan Sontag in her 1977 book *On Photography*. The first is that photographs cannot help but 'aestheticize' their subjects, and thus to grant them value, whether they have value or not. A book can express the low value of a subject, but a photograph struggles not to beautify. This is the camera's great ability, but it is likewise the reason it is so limited. Nevertheless, what the photograph did for the painting, allowing it to travel to the viewer rather than the other way around,[124] does perhaps echo what the arrival of movable-type printing on paper did for the book, allowing it to travel to the reader (or, at least, to many more readers than ever before).

The word-reader has an advantage over the film-viewer too because, whereas the reader can dictate his or her pace, the film-viewer is bound to the speed that the film dictates. The reader has greater freedom to focus and to skim, to evaluate and to dismiss. The viewer, especially in a cinema but in most other circumstances where he is not watching alone, has the option only to watch, to daydream, to sleep or to leave. In terms of how they communicate, films are far more directive of their audience than books. Of course, that makes them an easier form of escapism and often a preferred form of relaxation, but it does mean that books favour audience involvement and participation more than films do. They are a more interactive form, granting the reader greater control over the location and time-span of their experience. Films can present on their own terms, using visual 'facts' that are not easily dismissed, whereas books, which operate through an abstract intermediary – a printed script – have to work harder to persuade.

Worded paper has many mundane but crucial forms too, like flyers, tickets and posters. It has also found particular forms for particular pleasures, some only very late in its journey. A few of these have been phenomenally successful, among them comics. Thus the best-selling French authors abroad have not been Descartes, Voltaire and Balzac, but Goscinny and Uderzo, the creators of Asterix. (The *Asterix* comics have sold more than 320 million copies worldwide.) More recently still, the graphic novel has won both growing critical respect and an increasingly mainstream audience.

These idiosyncratic uses of paper are matched by its capacity for surprise – and for unintended consequences. Very few authors illustrate this more vividly than Martin Luther, who set out to reform Christian Europe, turning it back to the Bible as its source of authority, but who ended up not only dividing it but even contributing to the rise of European modernism, which he would have despised, both for its humanist self-belief and for its disregard of God. Paper's social impact has been so far-reaching that it has proved remarkably difficult to direct – and productive in ways none of its early adopters could have imagined. This unpredictability and versatility are, however, shared by its digital competitor – the ability to personalize what is delivered to you online (and the format it arrives in) each day is one of the digital-reader's enormous advantages.

One of the most surprising paper formats to survive for so long has been the broadsheet newspaper, although it seems, finally, to be on its way out. Growing up in London, it was easy to miss how strange it was to have people standing on platforms at train stations and under bus shelters – or even walking through the streets – reading a product almost as large as a car windscreen. Multiple headlines and a minimal page-count do not completely explain this cultural oddity. Even in its reduced Berliner, compact or tabloid formats, the physical newspaper struggles to rival the conveniences of digital alternatives. It cannot deliver news as quickly, or as compellingly, as the combined force of word, sound and moving images. The daily news package is still an excellent product, but it is almost as good on a screen and the ability to click on links makes it more interactive and easier to cross-reference. Eventually, even the word 'newspaper' could become an anachronism.

Kindles, iPads and Tablets should, by rights, replace the physical book. They are more convenient, offer thousands of books in one portable reading device, and owners can always add new books at any moment of the day or night. It means no need to visit bookshops – or to browse according to the bookshop's own classifications. It makes books cheaper as they simply transmit words directly to the reader. It has the potential to increase the role of the author, given that physical books are the product not of a single author but of a chain or network that includes agents, editors, authors, proofers, designers, illustrators, printers, bookbinders, distributors and bookshops. The book is very much a social product and very far from being an unalloyed dictation from author to audience. Shifting from the paper book to the screen has the potential to change that quite radically, cutting some of the intermediaries.

For the moment, the physical book still feels more authoritative, more the product of a validating chain of production from writer to publisher to printer to bookbinder to salesman. But even that is changing. New surfaces always take time to win authority in the eye of readers. Books trump e-readers for beauty, of course, and publishers have wisely invested in the appearance of their physical titles in recent years. Yet beauty is not why the book will survive. As for the bookshop and library, those visible landmarks which remind us of the presence and power of books, their most striking feature today is their casualty rate. It will not be bookshop nostalgia that will save the book either.

It is no longer the most effective store of knowledge, except perhaps for taking on sea voyages and mountain hikes, but the book will survive because it is self-contained and unalterable. One definition of civilization is an attempt at permanence and the book is certainly that. Other forms of storage, like CD-ROMs, change and subside and are never instantly accessible in the same way, at least not without electrical power. Moreover, they do not give physical form to whatever writings and images we choose to value – you can merely switch texts on the same device (and, potentially, be secretly viewed doing so by a hostile party from the other side of the world) – and the avoidance of physical form denies a text one of its most tangible tokens of importance.

The book, on the other hand, allows us not merely to know stories, ideas, arguments and poems, but to own them too. The book itself embodies them, ready for us to thumb them, to enjoy the fall of light on their plant-wrought papers, to annotate them, to lend them out, to grant them a place on our most prominent bookshelf or to dispatch them to a dusty corner.

Here I mean not any kind of book, like a scroll or a bamboo roll, both books of a kind, but the codex. The codex developed in the third and fourth centuries AD as a Christian phenomenon in a region not yet familiar with paper. It was intended to be varied in content; it was also, like Christianity, polyglot, and therefore open to learning not just from Greece and Rome but from other traditions too. It had a documentary emphasis (it suits regular cross-referencing) that would end up delivering to it an independent intellectual authority and, in time, freedom from official patronage. Its origins are not part of paper's story but, when paper arrived in the Middle East, it found an excellent partner in the codex. Moreover, its length meant that the codex was not just a piece of writing – it was a small personal library in its own right.[125]

During the Reformation, as Bible codices circulated throughout Europe, they helped to take the power to identify and define truth out of the controlling hands of an elite. That displacement effect is a feature of paper's story – in the Renaissance, Reformation, scientific revolution, Enlightenment, French Revolution and in the rise of universal suffrage and universal education. Each of these was a paper-made movement which sought to displace received wisdom, structural inequality or institutional authority – or all three.

The book is perhaps the greatest physical expression of this story. A screen may achieve what the book does in practical, informational terms; a car manual might as well be on a screen. But the screen will never carry the more final physical expression that makes the great virtue of the paper book so unique. It may not always deliver what is true, but it does deliver itself as a personal and durable item, thus encouraging the individual reader to critique it for him or for herself – and to keep it beyond a first reading, too.

Paper may not deliver all that its employer intends, as Martin Luther discovered, but it has nevertheless enabled unprecedented change. Paper may be limited by the cultural assumptions it represents, as with the Treaty of Waitangi, but it can also be used to question even itself. (Indeed, in the case of the Treaty, that is exactly what *has* happened.) Paper may face a very real threat to its use for text in the digital age, but it also has peculiar strengths as a cultural product, strengths which cannot all be replicated. Chief among these is its physical reception by readers as an independent item: a handheld extended piece of writing that can be physically owned.

It is in this sense that paper's greatest role has been as courier of books to individual owner-readers. What began in China more than two millennia ago continues to engage people in news, stories, poems and correspondence to this day. Paper cannot promise to deliver quality content, any more than it can always evade censorship or consistently refuse propaganda. It cannot even promise to deliver truth. But what it can do, it has done. And that, for untold billions, has been to place power in the reader's hands.

Notes

1. Marco Polo, *The Travels*, trans. Ronald Latham (Harmondsworth: Penguin, 1958), p. 147.
2. The year of Marco Polo's arrival in Beijing is widely debated, with 1274 and 1275 the two most widely accepted possibilities in recent years.
3. Or so it is estimated for Tolstoy, the editing of whose complete works (already at 90 volumes) is an ongoing project in Russia. Anon, 'Paperback Q & A: Rosamund Bartlett on *Tolstoy*', *Guardian*, 6 December 2011, accessed 30 July 2012, http://www.theguardian.co.uk/books/2011/dec/06/paperback-q-a-rosamund-bartlett-tolstoy.
4. Prior to this, paper had not dislodged the leaves used for writing, such as Talipat despite its use in diplomatic letters for several centuries.
5. Gustave Flaubert, *Madame Bovary*, trans. Eleanor Marx-Aveling (Ware: Wordsworth Editions, 1994), p. 146.
6. Despite the disagreement, I use 'writing' in this book to refer to writing that is broadly phonetic. I include in this designation scripts that can only represent syllables (and not individual sounds) as well as those that do not represent vowels.
7. William V. Harris, *Ancient Literacy* (Cambridge, Mass.: Harvard University Press, 1989), pp. 114, 167–73.
8. Paul Saenger, *Space Between Words: The Origins of Silent Reading* (Stanford, Calif.: Stanford University Press, 1997).
9. Li Feng, 'Literacy and the social contexts of writing in the Western Zhou', in *Writing and Literacy in Early China*, ed. Li Feng and David Prader Banner (Seattle: University of Washington Press, 2011), pp. 271–301.
10. Alan Chan, 'Laozi', *Stanford Encyclopedia of Philosophy*, 2 May 2013, accessed 20 September 2013, http://plato.stanford.edu/entries/laozi/.
11. Robin D. S. Yates, 'Soldiers, scribes and women: literacy among the lower orders in China', in *Writing and Literacy in Early China*, ed. Li

Feng and David Prader Banner (Seattle: University of Washington Press, 2011), pp. 339–69.

12. Gordon Barrass, *The Art of Calligraphy in Modern China* (Berkeley and Los Angeles: University of California Press, 2002), p. 20.

13. Tsuen-Hsuin Tsien, *Written on Bamboo and Silk* (Chicago and London: University of Chicago Press, 2004).

14. Richard Kurt Kraus, *Brushes With Power: Modern Politics and the Chinese Art of Calligraphy* (Berkeley and Los Angeles: University of California Press, 1991), p. 41.

15. Tsuen-Hsuin Tsien, *Collected Writings on Chinese Culture* (Hong Kong: The Chinese University of Hong Kong, 2011), p. 54.

16. Tsuen-Hsuin Tsien, *Written on Bamboo and Silk*, op. cit., p.145.

17. Ibid.

18. Eunuchs had been a staple of court life in China since the Zhou dynasty. Often forcibly castrated at a young age, eunuchs could work in the palace and rise to positions of significant influence. They were often portrayed as villains in the official histories.

19. Tsuen-Hsuin Tsien, *Written on Bamboo and Silk*, op. cit., p. 152.

20. Cited in Antje Richter, *Letters and Epistolary Culture in Early Medieval China* (Seattle: University of Washington Press, 2013), p. 31.

21. Herrlee Creel, *Studies in Early Chinese Culture* (London: Johnson Press, 1938).

22. Osip Mandelstam, 'The Egyptian Stamp', in *The Noise of Time* (Princeton: Princeton University Press, 1965), p. 133.

23. Aurel Stein, *On Ancient Central Asian Tracks* (London: Pantheon, 1941), p. 179.

24. Jeannette Mirsky, *Sir Aurel Stein: Archaeological Explorer* (Chicago: University of Chicago Press, 1998), pp. 4–5.

25. The online WestEgg Inflation Calculator (www.westegg.com/inflation), which uses official US CPI data to make its calculations, says that what cost you US$200 in 1910 would have cost you $4,620.52 in 2010.

26. Eric Zurcher, *The Buddhist Conquest of China: The Spread and Adaptation of Buddhism in Early Medieval China* (Leiden: Brill, 1959).

27. Tokiwa Daijo, in *Studies in Chinese Buddhism*, ed. Arthur F. Wright and Robert M. Somers (New Haven, Conn.: Yale University Press, 1990).

28. Lionel Giles, 'Dated Chinese manuscripts in the Stein Collection', *Bulletin of the School of Oriental and African Studies* 9 (4), 1939, pp. 1023–45.

29. Adapted from Zha Pingqiu, 'The substitution of paper for bamboo and the new trend of literary development in the Han, Wei and early Jin dynasties', *Frontiers of Literary Studies in China* 1 (1), 2007, pp. 26–49, DOI 10.1007/s11702-007-0002.

30. William T. Graham, 'Mi Heng's "Rhapsody on a Parrot"', *Harvard Journal of Asiatic Studies* 31 (9), 1979, pp. 39–54.

31. Mark Edward Lewis, *China Between Empires* (Cambridge, Mass.: Belknap Press, 2009), pp. 196–247.

32. Adapted from Zha Pingqiu, 'The substitution of paper for bamboo', op. cit., p. 40.

33. Ibid.

34. Ibid.

35. Ibid.

36. J. D. Schmidt, *Harmony Garden* (London: Routledge-Curzon, 2003), p. 98.

37. Endymion Wilkinson, *Chinese History: A Manual* (Cambridge, Mass.: Harvard University Press, 2000), p. 445.

38. There seems to be some confusion over the date, which cannot have been later than 404. Endymion Wilkinson (*Chinese History: A Manual*, p. 448) puts the date at 404, *The Cambridge History of Chinese Literature* (p. 201) at 402, and a third source places it in the fourth century.

39. Kang-i Sun Chang and Stephen Owen, *The Cambridge History of Chinese Literature: to 1375* (Cambridge: Cambridge University Press, 2010), p. 201.

40. Adapted from Delmer M. Brown, *The Cambridge History of Japan, Volume 1: Ancient Japan* (Cambridge: Cambridge University Press, 1993), p. 393.

41. Adapted from Richard Karl Payne, *Discourse and Ideology in Medieval Japanese Buddhism* (London: Routledge, 2006), p. 73.

42. Nicolas Bouvier, *L'Usage du monde* (Paris: Payot, 1952).

43. Quoted in Jean Elizabeth Ward, *Po Chu-i: A Homage* (Lulu.com, 2008), p. 4.

44. Charles Benn, *China's Golden Age. Everyday Life in the Tang Dynasty* (Oxford: Oxford University Press, 2002).

45. The translations of Bai Juyi's poetry in this chapter are the author's own, although heavily influenced by Arthur Waley, Newi Alley, Burton Watson and David Hinton.

46. Slaves were treated as property in Tang China.

47. Lothar Ledderose, *Ten Thousand Things: Module and Mass Production in Chinese Art* (Princeton, NJ: Princeton University Press, 2000).

48. Wu Shuling, 'The development of poetry helped by the ancient postal service in the Tang dynasty', *Frontiers of Literary Study in China* 4 (4), 2010, pp. 553–77.

49. Konrad Kessler, *Mani Forschungen über die manichäische Religion* (Berlin: G. Reimer, 1889), p. 336.

50. The tenth-century Arabic scholar Ibn al-Nadim listed one further work but conflated two of the list into one and omitted the psalms and prayers.

51. This is the view of Professor Desmond Durkin-Meisterernst, research co-ordinator for Turfan Studies at the Berlin-Brandenburg Academy of Sciences in Berlin (author's interview in 2011).

52. Adapted from Hans-Joachim Klimkeit, *Gnosis on the Silk Road* (San Francisco: HarperCollins, 1992), p. 139.

53. In his *Confessions*, Augustine wrote that as a Manichaean he had sought to please his lust and vanity, but that he would then travel to see the Manichaean priests to give them food so that they could release angels to liberate him and his friends from their material bondage, and from the evil of their actions.

54. Philip Schaff, *A Select Library of the Nicene and Post-Nicene Fathers of the Christian Church* (Grand Rapids, Mich.: Eerdmans, 1956), p. 206.

55. Glazed earthenware used for architectural purposes.

56. In fact, larger papers had already been made elsewhere. One tenth-century Chinese author wrote that papermakers in the eastern city of Huizhou had used the hold of a ship as a vat, with fifty workers lifting the sheet in time to the banging of a drum, before drying it over a large brazier to get rid of any lumps.

57. A. E. Cowley, 'The Samaritan Liturgy and Reading of the Law', *The Jewish Quarterly Review*, 7:1, October 1894, pp. 121–40 (p.123).

58. Gabriel Said Reynolds (ed.), *The Qur'an in its Historical Context* (Abingdon: Routledge, 2008), p. 15.

59. Christoph Luxenburg, *The Syro-Aramaic Reading of the Qur'an: A Contribution to the Decoding of the Language of the Koran* (Berlin: Verlag Hans Schiller, 2007).

60. Fred M. Donner, 'Islamic Furqan', *Journal of Semitic Studies* LII/2, Autumn 2007, pp. 279–300.

61. Ibid.

62. A colophon historically referred to a book's closing inscription and usually carried details of its chain of production; today the term is used for the publisher's emblem.

63. Olga Pinto, 'The libraries of the Arabs in the time of the Abbasids', in *Islamic Culture 3* (Hyderabad: Academic and Cultural Publications Charitable Trust, 1929), pp. 210–43.

64. Adapted from Annemarie Schimmel, *Calligraphy and Islamic Culture* (Albany, NY: State University of New York Press, 1984).

65. Jonathan Bloom, *Paper Before Print* (New Haven, Conn.: Yale University Press, 2001).

66. N. G. Wilson, 'The history of the book in Byzantium', in *The Oxford Companion to the Book*, ed. Michael F. Suarez and H. R. Woudhuysen (Oxford: Oxford University Press, 2010), p. 37.

67. S. M. Imamuddin, *Arab Writing and Arab Libraries* (London: Ta-Ha Publishers, 1983).

68. Mark Edwards, *Printing, Propaganda and Martin Luther* (Minneapolis: Augsburg Fortress Publishers, 2005), p. xii.

69. David Ganz, 'Carolingian manuscript culture and the making of the literary culture of the Middle Ages', in *Literary Cultures and the Material Book*, ed. Simon Eliot, Andrew Nash and Ian Willison (London: British Library Publishing, 2007), pp. 147–58.

70. Jonathan Bloom, *Paper Before Print*, op. cit.

71. Robert Burns, 'Paper comes to the West', in Uta Lindgren, *Europäische Technik im Mittelalter:800 bis 1400. Tradition und Innovation* (Berlin: Gebr. Mann Verlag, 1996), pp. 413–22.

72. Simon Eliot and Jonathan Rose, *A Companion to the History of the Book* (Oxford: Wiley-Blackwell, 2009), pp. 207–31.

73. Petrarch had the largest private classical library in fourteenth-century Europe. On one occasion, when he dropped a book by his beloved Cicero (a particular favourite with Renaissance readers more generally), he apologized to it before placing it back on his shelves above all the others.

74. Martin Luther, *Luther's Works*, trans. Gottfried G. Krodel, Vol. 48, *Letters* (Philadelphia: Fortress Press, 1963), pp. 12–13.

75. Martin Luther, *Luther's Works*, ed. Harold J. Grimm and Helmut T. Lehmann, Vol. 31, *Career of the Reformer I* (Philadelphia: Fortress Press, 1999 (first published 1957)), pp. 25–33.

76. Michael Mullett, *Martin Luther* (London and New York: Routledge, 2004), pp. 67–74.

77. Mark Edwards, *Printing, Propaganda and Martin Luther*, op. cit., pp.16.

78. Ibid., pp. 107.

79. Universial Short Title Catalogue, www.ustc.ac.uk, accessed 5 March 2014.

80. Mark Edwards, *Printing, Propaganda and Martin Luther*, op. cit., p.29.

81. Ibid., pp. 14–40.

82. Robert Scribner, *For the Sake of Simple Folk: Popular Propaganda for the German Reformation* (Oxford and New York: Oxford University Press, 1994).

83. Elizabeth Eisenstein, *The Printing Press as an Agent of Change* (Cambridge: Cambridge University Press, 1980), p. 131.

84. Diarmaid MacCulloch, *Reformation: Europe's House Divided, 1490–1700* (London: Penguin, 2004), p. 198.

85. Augustino Scarpinello to Francesco Sforza, Duke of Milan. From *Venice: December 1530, Calendar of State Papers Relating to English Affairs in the Archives of Venice, Volume 4: 1527–1533* (1871), pp. 265–73, http://www.british-history.ac.uk/report. aspx?compid=94613, accessed 20 June 2013.

86. David Daniell, *William Tyndale: A Biography* (New Haven, Conn.: Yale University Press, 2001).

87. Kari Konkola and Diarmaid MacCulloch, 'People of the Book: the success of the Reformation', *History Today*, October 2003, 53 (10).

88. Jean-François Gilmont, ed., *The Reformation and the Book* (Aldershot: Ashgate, 1998).

89. The standing of these books throughout Church history had never equalled that of the Bible's other books.

90. Adrian Johns, *The Nature of the Book: Print and Knowledge in the Making* (Chicago: University of Chicago Press, 2000), p. 42.

91. Lucien Febvre and Henri-Jean Martin, *The Coming of the Book: The Impact of Printing, 1450–1800* (London: Verso, 2010), pp. 167–215.

92. Hans Belting, *Florence and Baghdad: Renaissance Art and Arab Science* (Cambridge, Mass.: Harvard University Press, 2011), esp. pp. 90–99.

93. Gerald P. Tyson and Sylvia Stoler Wagenheim, *Print and Culture in the Renaissance: Essays on the Advent of Printing in Europe* (Newark, NJ: University of Delaware Press, 1986), pp. 222–45.

94. The rise of opera and orchestral music would demand multiple staves; the arrival of the pianoforte in the eighteenth century also made single-impression printing less practical. The next key innovation would be mosaic printing, invented in Vienna in the 1750s by Breitkopf, the world's oldest music publisher. This technique could be remarkably complex, requiring some 500 different type sorts – separate type heads were used for the note-head, stem and staff of each note.

95. Martin Luther, *Luther's Works*, ed. J. J. Pelikan, H. C. Oswald and H. T. Lehmann, Vol. 49 (Philadelphia: Fortress Press, 1972), pp. 427–8.

96. Piero Weiss and Richard Taruskin, *Music in the Western World: A History in Documents* (New York: Collier-Macmillan, 1984), pp. 135–7.

97. Andrew Pettegree, *The Book in the Renaissance* (New Haven, Conn. and London: Yale University Press, 2010), pp. 172–3.

98. Iain Fenlon, 'Music, print and society', in *European Music 1520–1640*, ed. James Haar (Woodbridge: The Boydell Press, 2006), p. 287.

99. Christoph Wolff, *Bach: Essays on his Life and Music* (Cambridge, Mass.: Harvard University Press, 1991), p. 93.

100. Terri Bourus, *Shakespeare and the London Publishing Environment: The Publisher and Printers of Q1 and Q2 Hamlet*, AEB, Analytical & Enumerative Bibliography 12 (DeKalb, Ill.: Bibliographical Society of Northern Illinois, 2001), pp. 206–22.

101. Anthony J. Cascardi, *The Cambridge Companion to Cervantes* (Cambridge: Cambridge University Press, 2002), p. 59.

102. Terry Eagleton, *The English Novel: An Introduction* (Oxford: Blackwell, 2005), p. 3.

103. Catherine M. Bauschatz, 'To choose ink and pen: French Renaissance women's writing', in *A History of Women's Writing in France*, ed. Sonya Stephens (Cambridge: Cambridge University Press, 2000), p. 47.

104. Terry Eagleton, *The English Novel*, op. cit., p. 8.

105. Charles Taylor, *The Sources of the Self* (Cambridge, Mass.: Harvard University Press, 2009), p. 287.

106. S. Hull, *Chaste, Silent and Obedient: English Books for Women 1475–1640* (San Marino, Calif.: Huntingdon Library, 1982).

107. Belinda Elizabeth Jack, *The Woman Reader* (New Haven, Conn.: Yale University Press, 2012), p. 265.

108. Peter L. Shillingsburg, *From Gutenberg to Google: Electronic Representations of Literary Texts* (Cambridge: Cambridge University Press, 2006), p. 128, citing Gordon N. Ray, *Bibliographical Resource for the Study of Nineteenth Century Fiction* (Los Angeles: Clark Library, 1964) and John Sutherland, 'Victorian novelists: who were they?' in *Victorian Writers, Publishers, Readers* (New York: St Martin's Press, 1995).

109. Ann Blair, 'The rise of note-taking in early modern Europe', *Intellectual History Review* 20 (3), 2010, pp. 303–16.

110. Paul Marcus Dover, 'Deciphering the archives of fifteenth-century Italy', *Archival Science* 7, 2007, p. 299.

111. Ibid.

112. Jacob Soll, *Publishing* The Prince: *History, Reading, and the Birth of Political Criticism* (Ann Arbor: University of Michigan Press, 2008).

113. *Bristol Mercury*, 2 August 1712.

114. Jeroen Blaak, *Literacy in Everyday Life: Reading and Writing in Early Modern Dutch Diaries*, trans. Beverley Jackson (Leiden: Koninklijke Brill, 2009), pp. 222–34.

115. S. Hart, *Geschrift en Getal* (Dordrecht: Historische Vereniging, 1976).

116. Elizabeth Eisenstein, '*Steal this Film*' interview, 'Steal this film' website, Washington DC, April, 2007, http://footage.stealthisfilm.com/video/4.

117. Robert Service, *Stalin. A Biography* (London: Macmillan, 2004), p. 10.

118. Robert Darnton, *The Forbidden Bestsellers of Pre-revolutionary France* (New York: Norton, 1995), p. 21.

119. Robert Darnton, *Revolution in Print* (Berkeley: University of California Press, 1989), pp. 91–3.

120. Carla Hesse, *Publishing and Cultural Politics in Revolutionary Paris, 1789–1810* (Berkeley: University of California Press, 1991), p. 167.

121. Albert Furtwangler, *The Authority of Publius: A Reading of the Federalist Papers* (Ithaca, NY: Cornell University Press, 1984), pp. 87–93.

122. *Massachusetts Gazette*, 13 November 1787, p. 3.

123. Judith Binney, 'Maori oral narratives and Pakeha texts: two ways of telling history', *New Zealand Journal of History* 27 (1), 2007, pp. 16–28.

124. John Berger, *Ways of Seeing* (Harmondsworth: Penguin, 1972).

125. Anthony Grafton and Megan Williams, *Christianity and the Transformation of the Book* (Cambridge, Mass.: Harvard University Press, 2006), pp. 1–21.

Bibliography

*And further, by these, my son, be admonished: of making
many books there is no end; and much study is a weariness of
the flesh.*

Ecclesiastes 12:12

Alley, Rewi, *Bai Juyi: 200 Selected Poems* (Beijing: New World Press,
  1983).
Anderson, Benedict, *Imagined Communities* (London: Verso, 1991).
Asimov, M. S. and Bosworth, Clifford Edmund, *History of Civilisa-
  tions of Central Asia, Vol. 4: The Age of Achievement 750 to the
  end of the 15th Century* (Paris: UNESCO, 1999).
Avrin, Leila, *Scribes, Scripts and Books: The Book Arts from Antiquity to the Renaissance* (London: British Library, 1991).
Awwa, Salwa Muhammad, *Textual Relations in the Qur'an: Rele-
  vance, Coherence and Structure* (London: Routledge, 2006).
Al-Azami, Muhammad Mastafa, *The History of the Qur'anic
  Texts: From Revelation to Compilation: A Comparative Study with
  the Old and New Testaments* (Leicester: UK Islamic Academy,
  2003).

Bagley, Robert W., 'Anyang writing and the origins of the Chinese writing system', in *The First Writing*, ed. Stephen D. Houston (Cambridge: Cambridge University Press, 2004).

Baker, Colin F., *Qur'an Manuscripts: Calligraphy, Illumination, Design* (London: British Library, 2007).

Barnard, John, McKenzie, D. F. and Bell, Maureen, eds., *The Cambridge History of the Book in Britain, Vol. 4* (Cambridge: Cambridge University Press, 2002).

Barrass, Gordon S., *The Art of Calligraphy in Modern China* (Berkeley: University of California Press, 2002).

Barrett, T. H., *The Woman who Discovered Printing* (New Haven, Conn. and London: Yale University Press, 2008).

— 'Stupa, sutra and sarira in China c.656–706', *Buddhist Studies Review* 18 (1), 2001, pp. 1–64.

— *Singular Listlessness: A Short History of Chinese Books and British Scholars* (London: Wellsweep, 1989).

— *Japanese Papermaking: Traditions, Tools and Techniques* (New York: Weatherhill, 1983).

Basbanes, Nicholas, *A Gentle Madness: Bibliophiles, Bibliomanes and the Eternal Passion for Books* (New York: Henry Holt, 1999).

Bauschatz, Catherine M., 'To choose ink and pen: French Renaissance women's writing', in *A History of Women's Writing in France*, ed. Sonya Stephens (Cambridge: Cambridge University Press, 2000).

Bekker-Nielsen, Hans, Sorenson, Bengt Algot and Borch, Marianne, eds., *From Script to Book: A Symposium* (Odense: Odense University Press, 1986).

Belting, Hans, *Florence and Baghdad: Renaissance Art and Arab Science* (Cambridge, Mass.: Harvard University Press, 2011).

Benn, Charles, *China's Golden Age: Everyday Life in the Tang* (Oxford and New York: Oxford University Press, 2004).

Berger, John, *Ways of Seeing* (London: Penguin, 1972).

Bielenstein, Hans, 'Lo-yang in later Han times', *Bulletin of the Museum of Far Eastern Antiquities* 48, 1976.

Binney, Judith, 'Maori oral narratives and Pakeha texts: two ways of telling history', *New Zealand Journal of History* 27 (1), 2007, pp. 16–28.

Blaak, Jeroen, *Literacy in Everyday Life: Reading and Writing in Early Modern Dutch Diaries*, trans. Beverley Jackson (Leiden: Koninklijke Brill, 2009).

Blair, Ann, 'The rise of note-taking in early modern Europe', *Intellectual History Review* 20 (3), 2010, pp. 303–16.

Blair, Sheila S., *The Art and Architecture of Islam* (New Haven, Conn.: Yale University Press, 1996).

Bloom, Jonathan M., *Paper Before Print: The History and Impact of Paper in the Islamic World* (New Haven, Conn.: Yale University Press, 2001).

— 'Revolution by the ream: a history of paper', *Saudi Aramco World* 50 (3), 1999, pp. 26–39.

— *The Art and Architecture of Islam: 1250–1800* (New Haven, Conn.: Yale University Press, 1994).

Bol, Peter K., 'Seeking common ground: Han literati under Jurchen rule', *Harvard Journal of Asiatic Studies* 47 (2), 1987, pp. 461–538.

Borges, Jorge Luis, *The Library of Babel*, trans. Andrew Hurley (Jaffrey, NH: David R. Godine, 2000).

Boureau, Alain and Chartier, Roger, *The Culture of Print* (Cambridge: Polity Press, 1989).

Bourus, Terri, *Shakespeare and the London Publishing Environment: The Publisher and Printers of Q1 and Q2 Hamlet*, AEB, Analytical & Enumerative Bibliography 12 (DeKalb, Ill.: Bibliographical Society of Northern Illinois, 2001).

Bouvier, Nicolas, *L'Usage du monde* (Paris: Payot, 1952).

Bowman, Alan K. and Woolf, Greg, eds., *Literacy and Power in the Ancient World* (Cambridge: Cambridge University Press, 1994).

Boylan, Patrick, *Thoth: The Hermes of Egypt* (Oxford: Oxford University Press, 1922).

Brokaw, Cynthia Joanne and Kai-Wing Chow, *Printing and Book Culture in Late Imperial China* (Berkeley: University of California Press, 2005).

Bronkhorst, Johannes, *Buddhist Teaching in India* (Boston: Wisdom Publications, 2009).

Brooks, Douglas A., *From Playhouse to Printing House: Drama and Authorship in Early Modern England* (Cambridge: Cambridge University Press, 2000).

Brown, Delmer M., *The Cambridge History of Japan, Volume 1: Ancient Japan* (Cambridge: Cambridge University Press, 1993).

Burns, Robert, 'Paper comes to the West', in *Europäische Technik im Mittelalter 800 bis 1400. Tradition und Innovation*, ed. Uta Lindgren (Berlin: Gebr. Mann Verlag, 1996), pp. 413–22.

Carter, Thomas Francis, *The Invention of Printing in China and its Spread Westward* (New York: Ronald Press, 1955).

Cascardi, Anthony J., *The Cambridge Companion to Cervantes* (Cambridge: Cambridge University Press, 2002).

Cavallo, Guglielmio and Chartier, Roger, *A History of Reading in the West* (Oxford: Polity Press, 1999).

Chaffee, John William, *The Thorny Gates of Learning in Sung China: A Social History of Examinations* (Cambridge: Cambridge University Press, 1985).

Chan, Alan, 'Laozi', *Stanford Encyclopaedia of Philosophy*, 2 May 2013, accessed 20 September 2013, http://plato.stanford.edu/entries/laozi/.

Chang, Kang-i Sun and Owen, Stephen, *The Cambridge History of Chinese Literature* (Cambridge: Cambridge University Press, 2010).

Chappell, David W., 'Hermeneutical phases in Chinese Buddhism', in *Buddhist Hermeneutics*, ed. Donald Lopez (Honolulu: University of Hawaii Press, 1988).

Chartier, Roger, *Cultural History: Between Practices and Representations* (Cambridge: Polity Press, 1988).

Chen Junpu, ed. and trans., *150 Chinese–English Quatrains by Tang Poets* [bilingual edition] (Shanghai: Shanghai University Press, 2005).

Chia, Lucille and Idema, W. L., *Books in Numbers* (Cambridge, Mass.: Harvard University Press, 2007).

Chibbett, David, *The History of Japanese Printing and Book Illustration* (Tokyo and New York: Kodansha International, 1977).

Chin, Annping, *The Authentic Confucius* (New York: Scribner, 2007).

Chow Kai-wing, *Publishing, Culture and Power in Early Modern China* (Stanford, Calif.: Stanford University Press, 2004).

Chun Fang Yü, *Kuan-yin: The Chinese Transformation of Avaloikitesvara* (New York: Columbia University Press, 2001).

Chun Shin-yong, *Buddhist Culture in Korea* (Seoul: International Culture Foundation, 1974).

Clanchy, M. T., *From Memory to Written Record: England 1066–1307* (Oxford: Basil Blackwell, 1993).

Cleaves, Francis Woodman, trans., *The Secret History of the Mongols* (Cambridge, Mass.: Harvard University Press, 1982).

Cole, Richard, 'The Reformation in print: German pamphlets and propaganda', *Archiv für Reformationsgeschichte* 66, 1975, pp. 93–102.

Cook, Michael, *The Koran: A Very Short Introduction* (Oxford: Oxford University Press, 2000).

Creel, Herrlee, *Studies in Early Chinese Culture* (London: Johnson Press, 1938).

Crone, Patricia and Hinds, Martin, *God's Caliph: Religious Authority in the First Centuries of Islam* (Cambridge: Cambridge University Press, 2003).

Daftary, Farhad, *Intellectual Traditions in Islam* (London and New York: I. B. Tauris, 2000).

Dane, Joseph, *The Myth of Print Culture* (Toronto: University of Toronto Press, 2003).

Daniell, David, *William Tyndale: A Biography* (New Haven, Conn.: Yale University Press, 2001).

Daniels, Peter, *The World's Writing Systems* (Oxford: Oxford University Press, 2010).

Darnton, Robert, *The Forbidden Bestsellers of Pre-Revolutionary France* (New York: Norton, 1995).

— *Revolution in Print* (Berkeley: University of California Press, 1989).

Dickens, A. G., *The English Reformation* (University Park, Pa.: The Pennsylvania State University Press, 1989).

Diringer, David, *The Book Before Printing: Ancient, Medieval, Oriental* (New York: Dover, 1982).

— *Writing* (London: Thames & Hudson, 1962).

Dover, Paul Marcus, 'Deciphering the archives of fifteenth century Italy', *Archival Science* 7, 2007, pp. 297–316.

Dubs, Homer, ed., *History of the Former Han Dynasty by Ban Gu* (London: Trübner, 1944).

Duffy, Eamon, *The Stripping of the Altars: Traditional Religion in England c1400–c1580* (New Haven, Conn. and London: Yale University Press, 2005).

Dumoulin, Heinrich, *Zen Buddhism: A History* (New York and London: Macmillan, 1988).

Durkin, Desmond, *Mani's Psalms: Middle Persian, Parthian and Sogdian Texts in the Turfan Collection* (Turnhout: Brepols, 2010).

— *Turfan Revisited: The First Century of Research into the Arts and Cultures of the Silk Road* (Berlin: Reimer, 2004).

Eagleton, Terry, *The English Novel: An Introduction* (Oxford: Blackwell, 2005).

Eco, Umberto and de la Carrière, Jean, *This is Not the End of the Book*, trans. Polly Mclean (London: Harvill Secker, 2011).

Edkins, Joseph, *Chinese Buddhism: A Volume of Sketches, Historical, Descriptive and Critical* (London: Routledge, 2000).

Edwards, Mark, *Printing, Propaganda and Martin Luther* (Minneapolis, Minn.: Fortress Press, 2005).

Eisenstein, Elizabeth, 'Steal this Film' interview, 'Steal this Film' website, Washington DC, April, 2007, http://footage.stealthisfilm.com/video/4, accessed July 2013.

— *The Printing Revolution in Early Modern Europe* (Cambridge: Cambridge University Press, 1983).

— *The Printing Press as an Agent of Change: Communications and Cultural Transformations in Early Modern Europe* (Cambridge: Cambridge University Press, 1980).

Eliot, Simon and Rose, Jonathan, *A Companion to the History of the Book* (Oxford: Wiley-Blackwell, 2009).

Esack, Farid, *The Quran: A User's Guide* (Oxford: One World, 2005).

Ettinghausen, Richard, Grabar, Oleg and Jenkins-Madina, Marilyn, *Islamic Art and Architecture 650–1200* (New Haven, Conn.: Yale University Press, 2003).

Farale, Dominique, *Les batailles de la région du Talas et l'expansion musulmane en Asie Centrale* (Paris: Economica, 2006).

Febvre, Lucien and Martin, Henri-Jean, *The Coming of the Book: The Impact of Printing, 1450–1800* (London: Verso, 2010).

Fenlon, Iain, 'Music, print and society', in *European Music 1520–1640*, ed. James Haar (Woodbridge: The Boydell Press, 2006).

Fierro, Maribel, ed., *The New Cambridge History of Islam, Vol. 2: The Western Islamic World, Eleventh to Eighteenth Centuries* (Cambridge: Cambridge University Press, 2010).

Finkelstein, David, *An Introduction to Book History* (New York and London: Routledge, 2005).

Fischer, Stephen Roger, *A History of Writing* (London: Reaktion, 2001).

Flaubert, Gustave, *Madame Bovary*, trans. Eleanor Marx-Aveling (Ware: Wordsworth Editions, 1994).

Franke, Herbert, *China Under Mongol Rule* (Aldershot: Variorum, 1994).

Frishman, Martin, *The Mosque: History, Architectural Development and Regional Diversity* (London: Thames & Hudson, 2002).

Frye, Richard N., *The Golden Age of Persia* (London: Weidenfeld & Nicolson, 1975).

Furtwangler, Albert, *The Authority of Publius: A Reading of the Federalist Papers* (Ithaca, NY: Cornell University Press, 1984).

Ganz, David, *Corbie in the Carolingian Renaissance* (Sigmaringen: Thorbecke, 1990).

Gawthrop, Richard and Strauss, Gerald, 'Protestantism and literacy in early modern Germany', *Past and Present* 104, 1984, pp. 31–55.

Gee, Malcolm and Kirk, Tim, *Printed Matters: Printing, Publishing and Urban Culture in Europe in the Modern Period* (Aldershot: Ashgate, 2002).

Gernet, Jacques, *Daily Life in China on the Eve of the Mongol Invasion 1250–1276*, trans. H. M. Wright (Stanford, Calif.: Stanford University Press, 1962).

Gibb, H. A. R., *Studies on the Civilization of Islam* (Princeton, NJ: Princeton University Press, 1982).

Giles, Lionel, 'Dated Chinese manuscripts in the Stein collection', *Bulletin of the School of Oriental Studies, University of London* 10 (2), 1940, pp. 317–44.

— 'Dated Chinese manuscripts in the Stein collection', *Bulletin of the School of Oriental and African Studies* 9 (4), 1939, pp. 1023–45.

Gilmont, Jean-François, *John Calvin and the Printed Book* (Kirksville, Mo.: Truman State University Press, 2005).

— ed., *The Reformation and the Book* (Aldershot: Ashgate, 1998).

Gode, P. K., 'Migration of paper from China to India', *Studies in Indian Cultural History* 3, 1964.

Golombek, Lisa, *Timurid Art and Culture: Iran and Central Asia in the Fifteenth Century* (Leiden: Brill, 1992).

Goody, Jack, *The Power of the Written Tradition* (Washington, DC: Smithsonian Institution, 2000).

— 'The consequences of literacy', *Comparative Studies in Society and History* 5(3), 1963, pp. 304–45.

Grafton, Anthony and Williams, Megan, *Christianity and the Transformation of the Book* (Cambridge, Mass.: Harvard University Press, 2006).

Graham, William T., 'Mi Heng's "Rhapsody on a Parrot"', *Harvard Journal of Asiatic Studies* 31 (9), 1979, pp. 39–54.

Green, V. H. H., *Renaissance and Reformation: A Survey of European History between 1440 and 1660* (London: Edward Arnold, 1964).

Griffiths, Dennis, *Fleet Street: Five Hundred Years of the Press* (London: British Library Publishing, 2006).

Grousset, René , *L'Empire des Steppes* (Paris: Payot, 2001).

Grudem, Wayne and Dennis, Lane T., eds., *English Standard Version Study Bible* (Wheaton, Ill.: Crossway Bibles, 2008).

Gulacsi, Zuzsanna, *Mediaeval Manichaean Book Art: A Codicological Study of Iranian and Turkic Illuminated Book Fragments [Nag Hammadi and Manichaean Studies]* (Leiden: Brill Academic, 2005).

Habein, Yaeko Sato, *History of the Japanese Written Language* (Tokyo: Tokyo University Press, 1984).

Haldhar, S. M., *Buddhism in China and Japan* (New Delhi: Om Publications, 2005).

Hannawi, Abdul Ahad, 'The role of the Arabs in the introduction of paper into Europe', *Middle Eastern Library Association* 85, 2012, pp. 14–29.

Harris, Roy, *The Origin of Writing* (London: Duckworth, 1986).

Harris, William V., *Ancient Literacy* (Cambridge, Mass.: Harvard University Press, 1989).

Hart, S., *Geschrift en Getal* (Dordrecht: HistorischeVereniging, 1976).

Heck, Paul L., *The Construction of Knowledge in Islamic Civilization: Qudama B. Ja'Far and His Kitab Al-Kharaj Wa-Sina'at Al-Kitaba* (Leiden: Brill, 2003).

Herman, Ann. *The Spread of Buddhism* (Leiden: Brill, 2007).

Heuser, Manfred and Klimkeit, Hans-Joachim, *Studies in Manichaean Literature and Art* (Leiden: Brill, 1998).

Hillier, Jack, *The Art of the Japanese Book* (London: Philip Wilson, 2007).

Hinton, David, ed. and trans., *The Selected Poems of Po Chü-i* (London: Anvil Press, 2006).

Hoang, Michael, *Genghis Khan* (London: Saqi, 1990).

Hoberman, Barry, 'The Battle of Talas', *Saudi Aramco World* September/October, 1982, pp. 26–31.

Hodgson, Marshall G. S., *The Venture of Islam, Volume One: The Classical Age of Islam* (Chicago and London: University of Chicago Press, 1977).

Hoernle, A. F. Rudolf, 'Who was the inventor of rag paper?' *Journal of the Royal Asiatic Society of Great Britain and Northern Ireland* 1903, pp. 663–84.

Hoggart, Richard, *The Uses of Literacy* (Harmondsworth: Penguin, 1957).

Hongkyung Kim, 'The original compilation of the Laozi: a contending theory on its Qin origin', *Journal of Chinese Philosophy* 34 (4), 2007, pp. 613–30.

Houston, Stephen D., ed.,*The First Writing* (Cambridge: Cambridge University Press, 2004).

Hull, S., *Chaste, Silent and Obedient: English Books for Women 1475–1640* (San Marino, Calif.: Huntingdon Library, 1982).

Hunter, Dard, *Papermaking: The History and Technique of an Ancient Craft* (New York: Dover, 1947).

Idema, Wilt and Haft, Lloyd, *A Guide to Chinese Literature* (Ann Arbor, Mich.: Center for Chinese Studies, University of Michigan, 1997).

Imamuddin, S. M., *Arab Writing and Arab Libraries* (London: Ta-Ha Publishers, 1983).

Ivanhoe, Philip J., *The Daodejing of Laozi* (Indianapolis and Cambridge: Hackett, 2002).

Jack, Belinda Elizabeth, *The Woman Reader* (New Haven, Conn.: Yale University Press, 2012).

Jasnow, Richard and Zauzich, Karl-Theodor, *The Ancient Egyptian Book of Thoth: A Demotic Discourse on Knowledge and Pendant to the Classical Hermetica* (Wiesbaden: Harassovitz Verlag, 2005).

Jayne, Sears Reynolds, *Library Catalogues of the Renaissance* (Berkeley: University of California Press, 1956).

Jeffery, Arthur, *Materials for the History of the Text of the Qur'an: The Old Codices* (Leiden: Brill, 1937).

Johns, Adrian, *The Nature of the Book: Print and Knowledge in the Making* (Chicago: University of Chicago Press, 2000).

Juvaini, Ata-Malik, *The History of the World-Conqueror*, trans. J. A. Boyle (Manchester: Manchester University Press, 1997).

Karabacek, J. and Baker, Don, *Arab Paper* (London: Archetype, 2007).

Kennedy, Hugh, *The Court of the Caliphs: The Rise and Fall of Islam's Greatest Dynasty* (London: Weidenfeld & Nicolson, 2004).

Kessler, Konrad, *Mani: Forschungen über die manichäische Religion* (Berlin: G. Reimer, 1889).

Khoo Seow Hwa and Penrose, Nancy L., *Behind the Brushstrokes: Appreciating Chinese Calligraphy* (Hong Kong: Asia 2000 Publishing, 1993).

Kim, H. G., 'Printing in Korea and its Impact on her Culture', MA dissertation, University of Chicago, 1973.

Klimkeit, Hans-Joachim, *Gnosis on the Silk Road* (San Francisco: HarperCollins, 1992).

Knobloch, Edgar, *Monuments of Central Asia* (London and New York: I. B. Tauris, 2001).

Kohn, Livia and Lafargue, Michael, *Lao-tzu and the Tao-te-Ching* (Albany, NY: State University of New York Press, 1998).

Konkola, Kari and MacCulloch, Diarmaid, 'People of the Book: the success of the Reformation', *History Today*, 53 (10), Oct. 2003.

Kornicki, Peter F., *The Book in Japan: A Cultural History from Beginnings to the Nineteenth Century* (Leiden: Brill, 1998).

Kramer, Samuel Noah, *History Begins at Sumer* (London: Thames & Hudson, 1981).

Kraus, Richard Kurt, *Brushes With Power: Modern Politics and the Chinese Art of Calligraphy* (Berkeley and Los Angeles: University of California Press, 1991).

Lai, T. C., *Treasures of a Chinese Studio: Ink, Brush, Inkstone, Paper* (Kowloon: Swindon Book Co., 1976).

Lancaster, Lewis R., *Introduction of Buddhism to Korea* (Berkeley: Asian Humanities Press, 1989).

— *The Korean Buddhist Canon, A Descriptive Catalogue* (Berkeley and London: University of California Press, 1979).

Lawrence, Bruce B., *The Quran: A Biography* (London: Atlantic Books, 2006).

Ledderose, Lothar, *Ten Thousand Things: Module and Mass Production in Art* (Princeton, NJ: Princeton University Press, 2000).

Ledyard, Gari Keith, *The Korean Language Reform of 1446* (Seoul: UMI Dissertation Services, 1998).

Lee, Peter, *Sourcebook of Korean Civilization* (New York: Columbia University Press, 1993).

Lee, Thomas H.C., 'Life in the schools of Sung China', *Journal of Asian Studies* 37 (1), 1977, pp. 45–60.

Legge, James, trans., *A Record of Buddhistic Kingdoms; Being an account by the Chinese monk Fa-Hien of his travels in India and Ceylon (A.D. 399–414) in search of the Buddhist Books of Discipline* (Oxford: Clarendon Press, 1886).

— trans., 'Bamboo Annals', in *The Chinese Classics* (London: Trübner, 1871).

Lester, Toby, 'What is the Koran?' *Atlantic Monthly* 283 (1), 1999.

Levi, Anthony, *Renaissance and Reformation: The Intellectual Genesis* (New Haven, Conn.: Yale University Press, 2002).

Lévy, André, *Chinese Literature, Ancient and Classical* (Bloomington: Indiana University Press, 2000).

Levy, Howard S., *Translations from Po Chü-i's Collected Works* (New York: Paragon, 1971).

Lewis, Mark Edward, *China Between Empires* (Cambridge, Mass.: Belknap Press, 2009).

— *Writing and Authority in Early China* (Albany: State University of New York Press, 1999).

Li Chi, *The Beginning of Chinese Civilization; Three Lectures Illustrated with Finds at An Yang* (Seattle: University of Washington Press, 1957).

Li Feng, 'Literacy and the social contexts of writing in the Western Zhou', in *Writing and Literacy in Early China*, ed. Li Feng and David Prader Banner (Seattle: University of Washington Press, 2011), pp. 271–301.

Lieu, Samuel N. C., *Manichaeism in China and Central Asia* (Leiden: Brill, 1998).

— *The Religion of Light: An Introduction to the History of Manichaeism in China* (Hong Kong: University of Hong Kong, 1979).

Lings, Martin and Safadi, Hassin, *The Qur'an: Catalogue of an Exhibition of Qur'an Manuscripts at the British Library, 3 April–15 August 1976* (London: World of Islam Publishing Co. for the British Library, 1976).

Lopez, Donald S. Jr., *Buddhism in Practice* (Princeton, NJ: Princeton University Press, 1995).

Loveday, Helen, *Islamic Paper: A Study of the Ancient Craft* (London: Archetype, 2007).

Lundbaek, Knud, 'The first translation from a Confucian classic in Europe', *China Mission Studies (1550–1800) Bulletin* 1, 1979, pp. 2–11.

Luo, Shubao, *An Illustrated History of Printing in China* (Hong Kong: Hong Kong University Press, 1998).

Lurie, David, 'The subterranean archives of early Japan: recently discovered sources for the study of writing and literacy', in *Books in Numbers*, ed. Lucille Chia and W. L. Idema (Cambridge, Mass.: Harvard University Press, 2007).

Luxenberg, Christoph, *The Syro-Aramaic Reading of the Qur'an: A Contribution to the Decoding of the Language of the Koran* (Berlin: Verlag Hans Schiller, 2007).

Lyons, Jonathan, *The House of Wisdom: How the Arabs Transformed Western Civilisation* (London: Bloomsbury, 2009).

Mabie, Hamilton Wright, *Norse Mythology: Great Stories from the Eddas* (New York: Dover, 2002).

McAuliffe, Jane Dammen, *The Cambridge Companion to the Quran* (Cambridge: Cambridge University Press, 2006).

MacCulloch, Diarmaid, *A History of Christianity* (London: Penguin, 2009).

— *Reformation: Europe's House Divided 1490–1700* (London: Penguin, 2004).

— *Thomas Cranmer: A Life* (New Haven, Conn.: Yale University Press, 1996).

McDermott, Joseph P., *A Social History of the Chinese Book: Books and Literati Culture in Late Imperial China* (Hong Kong: Hong Kong University Press, 2006).

McKenzie, Donald Francis, *Oral Culture, Literacy and Print in Early*

New Zealand: The Treaty of Waitangi (Wellington, New Zealand: Victoria University Press, 1985).

McMullen, David, *State and Scholars in Tang China* (Cambridge: Cambridge University Press, 1988).

McNair, Amy, *Donors of Longmen* (Honolulu: University of Hawaii Press, 2007).

Mair, Victor H., *The Shorter Colombia Anthology of Chinese Literature* (New York: Colombia University Press, 2000).

— 'Buddhism and the rise of the written vernacular in East Asia: the making of national languages', *Journal of Asian Studies* 53 (3), 1994, pp. 707–51.

— 'Script and language in medieval vernacular Sinitic', *Journal of the American Oriental Society* 112 (2), 1992, pp. 269–78.

— *T'ang Transformation Texts: A Study of the Buddhist Contribution to the Rise of Vernacular Fiction and Drama in China* (Cambridge, Mass.: Council on East Asian Studies, Harvard University, 1989).

— *Tun-huang Popular Narratives* (Cambridge: Cambridge University Press, 1983).

Mandelstam, Osip, 'The Egyptian stamp', in *The Noise of Time* (Princeton, NJ: Princeton University Press, 1965).

Manguel, Alberto, *A History of Reading* (London: HarperCollins, 1996).

Mann, Nicholas, 'Petrarca Philobiblon: the author and his books', in *Literary Cultures and the Material Book*, ed. Simon Eliot, Andrew Nash and Ian Willison (London: British Library Publishing, 2007).

Martin, Henri-Jean and Cochrane, Lydia G., *The History and Power of Writing* (Chicago: University of Chicago Press, 1994).

Mason, Haydn T., *The Darnton Debate: Books and Revolution in the Eighteenth Century* (Oxford: Voltaire Foundation, 1998).

Melton, James van Horn, *Cultures of Communication from Reformation to Enlightenment: Constructing Publics in the Early Modern German Lands* (Aldershot: Ashgate, 2002).

— *The Rise of the Public in Enlightenment Europe* (New York and Cambridge: Cambridge University Press, 2001).

Miller, Constance R., *Technical Prerequisite for the Invention of Printing in China and the West* (San Francisco: Chinese Materials Center, 1983).

Milton, John, *Areopagitica* (Indianapolis: Liberty Fund, 1999).

Mirsky, Jeannette, *Sir Aurel Stein: Archaeological Explorer* (Chicago: University of Chicago Press, 1998).

Miyazaki, Ichisada, *China's Examination Hell: The Civil Service Examinations of Imperial China*, trans. Conrad Schirokauer (New Haven, Conn. and London: Yale University Press, 1981).

Mollier, Christine, *Buddhism and Taoism Face to Face* (Honolulu: University of Hawaii Press, 2008).

Morgan, David, *The Mongols* (Oxford: Blackwell, 1990).

Müller, F. Max *The Sacred Books of the East* (Oxford: Clarendon Press, 1879–1910).

Mullett, Michael A., *Martin Luther* (London and New York: Routledge, 2004).

Myers, Robin, *The Stationers Company Archive: An Account of the Records 1554–1984* (Winchester: St Paul's, 1990).

Narain, A. K., *Studies in the History of Buddhism* (Delhi: BR Publishing, 2000).

Nasr, Seyyed Hossein, *Science and Civilization in Islam* (Chicago: ABC/Kazi, 2001).

Nattier, Jan, *A Guide to the Earliest Chinese Buddhist Translation* (Tokyo: Soka University, 2008).

Needham, Joseph, *Science and Civilisation in China*, 27 vols. (Cambridge: Cambridge University Press, 1954–2008).

Neuwirth, Angelika, Sinai, Nicolai and Marx, Michael, *The Qur'an in Context: Historical and Literary Investigations into the Qur'anic Milieu* (Leiden and Boston: Brill, 2010).

Noble, Richmond Samuel Howe, *Shakespeare's Biblical Knowledge: and the Use of the Book of Common Prayer as Exemplified in the Plays of the First Folio* (New York: Macmillan, 1935).

Olivelle, Patrick, trans., *The Law Code of Manu* (Oxford: Oxford University Press, 2004).

Payne, Richard Karl, *Discourse and Ideology in Medieval Japanese Buddhism* (London: Routledge, 2006).

Pelikan, Jaroslav, Oswald, H. C., Lehmann, H. T., Lundeen, Joel W. et al., eds., *Luther's Works*, 54 vols. (St Louis, Mo.: Concordia and Philadelphia: Fortress Press, 1955–86).

Peters, F. E., *The Voice, the Word, the Books: The Sacred Scriptures of the Jews, the Muslims, the Christians* (Princeton, NJ: Princeton University Press, 2007).

Pettegree, Andrew, *The Book in the Renaissance* (New Haven, Conn. and London: Yale University Press, 2011).

— *The French Book and the European World* (Leiden: Brill, 2007).

Pinto, Olga, 'The libraries of the Arabs during the time of the Abbasids', *Islamic Culture 3* (Hyderabad: Academic and Cultural Publications Charitable Trust, 1929), pp. 210–43.

Polastron, Lucien, *Books on Fire: The Tumultuous Stories of the World's Great Libraries* (London: Thames & Hudson, 2007).

Polo, Marco, *The Travels*, trans. Ronald Latham (Harmondsworth: Penguin, 1958).

Reeve, John, ed., *Sacred: Exhibition Catalogue* (London: British Library Publishing, 2007).

Reynolds, Gabriel Said, *The Qur'an in its Historical Context* (London and New York: Routledge, 2008).

Rezvan, E. A., 'The Quran and its world. VI: Emergence of the canon: the struggle for uniformity', *Manuscripta Orientalia* 4 (2), 1998, pp. 13–54.

Richardson, Brian F., 'The diffusion of literature in Renaissance Italy: the case of Pietro Bembo', in *Literary Cultures and the Material Book*, ed. Simon Eliot, Andrew Nash and Ian Willison (London: British Library Publishing, 2007), pp. 175–89.

— *Printing, Writers and Readers in Renaissance Italy* (Cambridge: Cambridge University Press, 1999).

Robinson, Chase F., ed., *The New Cambridge History of Islam, Vol. 1: The Formation of the Islamic World, Sixth to Eleventh Centuries* (Cambridge: Cambridge University Press, 2010).

— ed., *The New Cambridge History of Islam, Vol. 3: The Eastern Islamic World, Eleventh to Eighteenth Centuries* (Cambridge: Cambridge University Press, 2010).

Rogerson, Barnaby, *The Heirs of the Prophet Muhammad and the Roots of the Sunni–Shia Schism* (London: Little, Brown, 2006).

— *The Prophet Muhammad* (London: Little, Brown, 2003).

Saenger, Paul, 'Reading in the later Middle Ages', in *A History of*

Reading in the West, ed. Guglielmio Cavallo and Roger Chartier (Oxford: Polity Press, 1999), pp. 120–48.

— 'The history of reading', in *Literacy: An International Handbook*, ed. Daniel A. Wagner, Richard L. Vensky and Brian V. Street (Boulder, Colo.: Westview Press, 1999), pp. 11–15.

— *Space Between Words: The Origins of Silent Reading* (Stanford, Calif.: Stanford University Press, 1997).

Saheeh International, ed., *The Qur'an* (Riyadh: Abul Qaseem Publishing House, 1997).

Said, Labib, *The Recited Koran: A History of the First Recorded Version* (Princeton, NJ: Darwin, 1975).

Sanford, James H., LaFleur, William R. and Nagatomi, Masatoshi, *Flowing Traces: Buddhism in the Literary and Visual Arts of Japan* (Princeton NJ: Princeton University Press, 1992).

Sato, Masayuki, *The Confucian Quest for Order: The Origin and Formation of the Political Thought of Xunzi* (Leiden and Boston: Brill, 2003).

Schafer, Edward H., *The Golden Peaches of Samarkand: A Study of Tang Exotics* (Berkeley: University of California Press, 1963).

Schaff, Philip, *A Select Library of the Nicene and Post-Nicene Fathers of the Christian Church* (Grand Rapids, Mich.: Eerdmans, 1956).

Schimmel, Annemarie, *Calligraphy and Islamic Culture* (Albany, NY: State University of New York Press, 1984).

Schmidt, J. D., *Harmony Garden* (London: Routledge-Curzon, 2003).

Scribner, Robert, *For the Sake of Simple Folk: Popular Propaganda for the German Reformation* (Oxford and New York: Oxford University Press, 1994).

Sellman, James D., *Timing and Rulership in Master Lu's Spring and Autumn Annals* (Albany, NY: State University of New York Press, 2002).

Sells, Michael, *Approaching the Qur'an: The Early Revelations* (Ashland: White Cloud Press, 2007).

Service, Robert, *Stalin. A Biography* (London: Macmillan, 2004).

Sharpe, Kevin, *The Politics of Reading in Early Modern England* (New Haven, Conn.: Yale University Press, 2000).

Shaughnessy, Edward L., *Before Confucius: Studies in the Creation of the Chinese Classics* (Albany, NY: State University of New York Press, 1997).

— *Sources of Early Chinese History* (Berkeley, Calif.: Society for the Study of Early China and the Institute of East Asian Studies, 1997).

Sherman, William, *Used Books: Marking Readers in Renaissance England* (Philadelphia, Pa.: University of Pennsylvania Press, 2008).

Shillingsburg, Peter L., *From Gutenberg to Google: Electronic Representations of Literary Texts* (Cambridge: Cambridge University Press, 2006).

Sid, Muhammad Ata, *The Hermeneutical Problem of the Qur'an in Islamic History* (London: University Microfilms International, 1981).

Sima Qian, *Records of the Grand Historian, Han Dynasty I and II*, trans. Burton Watson (New York and Hong Kong: Columbia University Press, 1993).

Sinai, Nicolai, 'The Qur'an as process', in *The Qur'an in Context*, ed. Angelika Neuwirth, Nicolai Sinai and Michael Marx (Leiden and Boston: Brill, 2010).

Sinor, Denis, ed., *The Cambridge History of Early Inner Asia* (Cambridge: Cambridge University Press, 1990).

Slater, John Rothwell, *Printing and the Renaissance: A Paper Read before the Fortnightly Club of Rochester, New York* (New York: William Edwin Rudge, 1921).

Soll, Jacob, *Publishing* The Prince: *History, Reading, and the Birth of Political Criticism* (Ann Arbor, Mich.: University of Michigan Press, 2008).

Soucek, Svat, *A History of Inner Asia* (Cambridge: Cambridge University Press, 2000).

Spence, Jonathan D., *The Search for Modern China* (New York and London: W. W. Norton & Co., 1999).

Stein, Aurel, *On Ancient Central Asian Tracks* (London: Pantheon 1941).

Stock, Brian, *Augustine the Reader: Meditation, Self-knowledge and the Ethics of Interpretation* (Cambridge, Mass.: Harvard University Press, 1996).

Suarez, Michael F. and Woudhuysen, H. R., *The Oxford Companion to the Book, Vols. I & II* (Oxford: Oxford University Press, 2010).

Sugarman, Judith, 'Hand papermaking in China', *Hand Papermaking* 5 (1), 1990.

Sun Dayu, trans., *An Anthology of Ancient Chinese Poetry and Prose* (Shanghai: Shanghai Foreign Language Education Press, 1997).

Sutherland, John, 'Victorian novelists: who were they?' in *Victorian Writers, Publishers, Readers* (New York: St Martin's Press, 1995).

Tanaka, Kenneth, *The Dawn of Chinese Pure Land Buddhist Doctrine* (Albany, NY: State University of New York Press, 1990).

Taylor, Charles, *The Sources of the Self* (Cambridge, Mass.: Harvard University Press, 2009).

Thien An, *Buddhism and Zen in Vietnam in Relation to the Development of Buddhism in Asia* (Tokyo: Charles & Tuttle, 1975).

Thompson, Claudia, *Recycled Papers: The Essential Guide* (Boston, Mass.: MIT Press, 1992).

Tseng Yuho, *A History of Chinese Calligraphy* (Hong Kong: Chinese University Press, 1993).

Tsien Tsuen-hsuin, *Collected Writings on Chinese Culture* (Hong Kong: The Chinese University of Hong Kong, 2011).

— *Written on Bamboo and Silk: The Beginnings of Chinese Books and Inscriptions* (Chicago and London: University of Chicago Press, 2004, rev. ed. 2013).

Tsukamoto, Zenryu, *A History of Early Chinese Buddhism: From its Introduction to the Death of Hui-yüan* (New York: Kodansha International, 1985).

Tyson, Gerald P. and Wagenheim, Sylvia Stoler, *Print and Culture in the Renaissance: Essays on the Advent of Printing in Europe* (Newark, NJ: University of Delaware Press, 1986).

Vasari, Giorgio, *Lives of the Artists, Volume 1* (Harmondsworth: Penguin, 1965).

Venice: December 1530, Calendar of State Papers Relating to English Affairs in the Archives of Venice, Vol. 4: 1527–1533, Augustino Scarpinello to Francesco Sforza, Duke of Milan, pp. 265–73, http://www.british-history.ac.uk/report.aspx?compid=94613, accessed 20 June 2013.

Waley, Arthur, ed. and trans., *Chinese Poems* (London: George Allen and Unwin, 1956).

— *The Life and Times of Po Chü-i, 772–846 AD* (London: George Allen and Unwin, 1949).

Ward, Jean Elizabeth, *Po Chu-i: A Homage* (Lulu.com, 2008).

Warraq, Ibn, *Which Koran? Variants, Manuscripts, Linguistics* (Amherst, NY: Prometheus, 2011).

— *The Origins of the Koran* (Amherst, NY: Prometheus, 1998).

Watson, Burton, *The Columbia Book of Chinese Poetry: From Early Times to the Thirteenth Century* (New York: Columbia University Press, 1984).

Watson, Peter, *The German Genius: Europe's Third Renaissance, The Second Scientific Revolution and the Twentieth Century* (London: Simon & Schuster, 2010).

Weiss, Piero and Taruskin, Richard, *Music in the Western World: A History in Documents* (New York: Collier-Macmillan, 1984).

Welch, Theodore F., *Toshokan: Libraries in Japanese Society* (London: Clive Bingley, 1976).

Wells, Stanley and Taylor, Gary, eds., *The Oxford Shakespeare: The Complete Works* (Oxford: Oxford University Press, 1988).

Wessels, Anton, *Understanding the Qur'an* (London: ACM, 2001).

Westcott, W. W., *Sepher Yetzirah* (Cambridge: Milton, 1978, first published 1887).

Whitfield, Susan, *Life Along the Silk Road* (London: John Murray, 1999).

Wiet, Gaston, *Baghdad: Metropolis of the Abbasid Caliphate* (Oklahoma: University of Oklahoma Press, 1971).

Wild, Stefan, *The Quran as Text* (Leiden: Brill, 1997).

Wilkinson, Endymion, ed., *Chinese History: A Manual* (Cambridge, Mass.: Harvard University Press, 2000).

Willes, Margaret, *Reading Matters: Five Centuries of Discovering Books* (New Haven, Conn.: Yale University Press, 2008).

Wilson, Derek, *Out of the Storm: Life and Legacy of Martin Luther* (London: Pimlico, 2007).

Wolff, Christoph, *Bach: Essays on his Life and Music* (Cambridge, Mass.: Harvard University Press, 1991).

Woolf, Greg, 'Power and the spread of writing in the West', in *Literacy and Power in the Ancient World*, ed. Alan K. Bowman and Greg Woolf (Cambridge: Cambridge University Press, 1994), pp. 84–98.

Wright, Arthur F. and Somers, Robert M., eds., *Studies in Chinese Buddhism* (New Haven, Conn.: Yale University Press, 1990).

Wright, Arthur F. and Twitchett, Denis, *Confucian Personalities* (Stanford, Calif.: Stanford University Press, 1962).

Wu, K.T., 'Chinese printing under four alien dynasties: 916–1368 AD', *Harvard Journal of Asiatic Studies* 13 (3/4), 1950, pp. 447–523.

Wu Shuling, 'The development of poetry helped by ancient postal service in the Tang dynasty', *Frontiers of Literary Studies in China* 4 (4), 2010, pp. 553–77.

Wu Wei, trans., *The I Ching* (Los Angeles: Power Press, 2005).

Xiao, Gongquan and Mote, Frederick W., *A History of Chinese Political Thought: Volume 1: From the Beginnings to the Sixth Century BC* (Princeton, NJ: Princeton University Press, 1979).

Xiong, Victor Cunrui, *Sui-Tang Chang'an: A Study in the Urban History of Medieval China* (Ann Arbor, Mich.: Center for Chinese Studies, University of Michigan, 2000).

Xueqin Li, et al., 'The earliest writing? Sign use in the seventh millennium BC at Jiahu, Henan province, China', *Antiquity* 77 (295), 2003, pp. 31–44.

Yang Xuanzhi, *A Record of Buddhist Monasteries in Luoyang*, trans. Yitung Wang (Princeton, NJ: Princeton University Press, 1984).

Yates, Robin D. S., 'Soldiers, scribes and women: literacy among the lower orders in China', in *Writing and Literacy in Early China*, ed. Li Feng and David Prader Banner (Seattle: University of Washington Press, 2011), pp. 339–69.

Yu, Pauline, Bol, Peter, Owen, Stephen and Peterson, Willard, *Ways with Words: Writing About Reading Texts from Early China* (Berkeley: University of California Press, 2000).

Zha Pingqiu, 'The substitution of paper for bamboo and the new trend of literary development in the Han, Wei and early Jin dynasties', *Frontiers of Literary Studies in China* 1 (1), 2007, pp. 26–49.

Zürcher, Erik, 'Buddhism and education in Tang times', in *Neo-Confucian Education: The Formative Stage*, ed. William Theodore de Bary and John W. Chaffee (Berkeley: University of California Press, 1989).

— *The Buddhist Conquest of China: The Spread and Adaptation of Buddhism in Early Medieval China* (Leiden: Brill, 1959).

Illustration Credits

More detail about illustrations can be found below.

1. 'Spring' by Yang Xin. Owned by the author. Copyright applies.
2. Impression of cuneiform (by author).
3. Map by Jeff Edwards.
4. Youguan Temple (photograph by author).
5. Du Jin (fl. 1465–1509): *The Scholar Fu Sheng Transmitting the Book of Documents*, 15th/mid-16th century. China, Ming dynasty (1368–1644). New York, Metropolitan Museum of Art. Hanging scroll; ink and colour on silk. Image: 57 7/8 x 41 1/8"(147 x 104.5 cm). Overall with rollers: 10' 6" x 53" (320 x 134.6 cm. Gift of Douglas Dillon, 1991 (1991.117.2). © 2013. Image copyright The Metropolitan Museum of Art/Art Resource/Scala, Florence.
6. 'Happy' by Yang Xin. Owned by the author. Copyright applies.
7. The maceration process in central China (photograph by Leone Nani), © Archivio PIME.
8. Map by Jeff Edwards.
9. *Paul Pelliot dans la 'niche aux manuscrits' dans la grotte 17 de Mogao*, Mission Paul Pelliot 1906/1908, AP8186; AP8187, Pelliot Paul (1878–1945) © Droits reserves Pelliot Paul (1878–1945) © Droits reserves. Localisation: Paris, musée Guimet-musée national des Arts asiatiques © Musée Guimet, Paris, Dist. RMN-Grand Palais / Thierry Ollivier.
10. Traditionally attributed to: Yan Liben, Chinese, about 600–673. *Northern Qi scholars collating classic texts* (detail). Chinese, Northern Song dynasty, 11th century. Ink and colour on silk.

Overall: 28.5 x 731.2 cm (11 ¼ x 287 7/8 in.) Image: 27.5 x 114 cm (10 13/16 x 44 7/8 in.) Museum of Fine Arts, Boston. Denman Waldo Ross Collection, 31.123.

11. Personal stamp of Yang Xin (detail from calligraphy owned by author).

12. Map by Jeff Edwards.

13. Friday Mosque (photograph by author).

14. Map by Jeff Edwards.

15. *Folios from a Qur'an Manuscript.* Islamic. Iran, Isfahan. Abbasid period (750–1258). New York, Metropolitan Museum of Art. Ink and gold on paper, 9 7/16 x 13 13/16 in. (24 x 35.1 cm); Mat: 14 1/4 x 19 1/4 in. (36.2 x 48.9 cm); Frame: 15 1/4 x 20¼ in. (38.7 x 51.4 cm). Rogers Fund, 1940 (40.164.5a, b). © 2013. Image copyright The Metropolitan Museum of Art/Art Resource/ Scala, Florence.

16. *Liberté de la Presse* (1797), © Bibliothèque Nationale de France (Liberté de la Presse ID/Cote: QB-1 (1797-01/1797-05)-FOL, M-103642 Recueil. Documents sur l'histoire de France. Janvier-mai 1797).

17. Pressing papers in central China (photograph by Leone Nani), © Archivio PIME.

Index

Figures in bold refer to illustrations or their captions